GRAPHING CALCULATOR MANUAL

JUDITH A. PENNA
Indiana University Purdue University Indianapolis

ALGEBRA AND TRIGONOMETRY GRAPHS & MODELS

AND

PRECALCULUS: GRAPHS & MODELS

FOURTH EDITION

Marvin L. Bittinger
Indiana University Purdue University Indianapolis

Judith A. Beecher
Indiana University Purdue University Indianapolis

David J. Ellenbogen
Community College of Vermont

Judith A. Penna
Indiana University Purdue University Indianapolis

PEARSON
Addison Wesley

Boston San Francisco New York
London Toronto Sydney Tokyo Singapore Madrid
Mexico City Munich Paris Cape Town Hong Kong Montreal

Reproduced by Pearson Addison-Wesley from electronic files supplied by the author.

Copyright © 2009 Pearson Education, Inc.
Publishing as Pearson Addison-Wesley, 75 Arlington Street, Boston, MA 02116.

ISBN-13: 978-0-321-53198-8
ISBN-10: 0-321-53198-1

4 5 6 BRR 10 09

Contents

The TI-83 Plus and TI-84 Plus Graphing Calculators

Chapter R
Basic Concepts of Algebra

This section of the Graphing Calculator Manual provides keystrokes and suggestions for using the TI-83 Plus, the TI-84 Plus, and the TI-84 Plus Silver Edition graphing calculators.

GETTING STARTED

Before turning on the calculator note that there are options above the keys as well as on them. To access the option written on a key, simply press the key. The options written to the left above the keys are accessed by first pressing the $\boxed{\text{2nd}}$ key in the left column of the keypad and then pressing the key corresponding to the desired option. These options and the $\boxed{\text{2nd}}$ key are yellow on the TI-83 Plus and are blue on the TI-84 models. The options written in green to the right above the keys are accessed by first pressing the green $\boxed{\text{ALPHA}}$ key in the left column of the keypad.

Press $\boxed{\text{ON}}$ to turn on the calculator. ($\boxed{\text{ON}}$ is the key at the bottom left-hand corner of the keypad.) You should see a blinking rectangle, or cursor, on the screen. If you do not see the cursor, try adjusting the display contrast. To do this, first press $\boxed{\text{2nd}}$. Then press and hold $\boxed{\triangle}$ to increase the contrast or $\boxed{\triangledown}$ to decrease the contrast. If the contrast needs to be adjusted further after the first adjustment, press $\boxed{\text{2nd}}$ again and then hold $\boxed{\triangle}$ or $\boxed{\triangledown}$ to increase or decrease the contrast, respectively.

To turn the calculator off, press $\boxed{\text{2nd}}$ $\boxed{\text{OFF}}$. (OFF is the second operation associated with the $\boxed{\text{ON}}$ key.) The calculator will turn itself off automatically after about five minutes without any activity.

Press $\boxed{\text{MODE}}$ to display the MODE settings. Initially you should select the settings on the left side of the display as shown below. The Mode screen on a TI-83 Plus is shown below. The Mode screen on a TI-84 Plus displays the same settings but uses a different font. A clock also appears at the bottom of the screen.

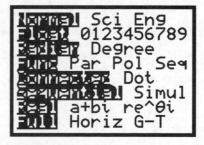

To change a setting on the Mode screen use $\boxed{\triangledown}$ or $\boxed{\triangle}$ to move the cursor to the line of that setting. Then use $\boxed{\triangleright}$ or $\boxed{\triangleleft}$ to move the blinking cursor to the desired setting and press $\boxed{\text{ENTER}}$. Press $\boxed{\text{CLEAR}}$ or $\boxed{\text{2nd}}$ $\boxed{\text{QUIT}}$ to leave the MODE screen. (QUIT is the second operation associated with the $\boxed{\text{MODE}}$ key.) Pressing $\boxed{\text{CLEAR}}$ or $\boxed{\text{2nd}}$ $\boxed{\text{QUIT}}$ will take you to the home screen where computations are performed.

It will be helpful to read the Getting Started section of the Texas Instruments Guidebook that was packaged with your graphing calculator before proceeding.

ABSOLUTE VALUE

Section R.1, Example 3 Find the distance between -2 and 3.

The distance between -2 and 3 is $|-2-3|$, or $|3-(-2)|$. Absolute value notation is denoted "abs" on the calculator. It is item 1 on the MATH NUM menu.

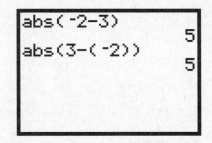

Note that the $\boxed{(-)}$ key in the bottom row of the keypad must be used to enter a negative number. The $\boxed{-}$ key in the right column of the keypad is used for the subtraction operation. Thus, to enter $|-2-3|$ press $\boxed{\text{MATH}}$ $\boxed{\triangleright}$ $\boxed{\text{ENTER}}$ $\boxed{(-)}$ $\boxed{2}$ $\boxed{-}$ $\boxed{3}$ $\boxed{)}$ $\boxed{\text{ENTER}}$. To enter $|3-(-2)|$ press $\boxed{\text{MATH}}$ $\boxed{\triangleright}$ $\boxed{\text{ENTER}}$ 3 $\boxed{-}$ $\boxed{(}$ $\boxed{(-)}$ $\boxed{2}$ $\boxed{)}$ $\boxed{)}$ $\boxed{\text{ENTER}}$. Note that the calculator supplies the left parenthesis in the absolute value notation. We close the expression with a right parenthesis although it is not necessary to do so. It is not necessary to include the inner set of parentheses around -2 in the second computation, but they allow the expression to be read more easily so we include them here.

```
abs(-2-3)
                    5
abs(3-(-2))
                    5
```

Instead of pressing $\boxed{\text{MATH}}$ $\boxed{\triangleright}$ $\boxed{\text{ENTER}}$ to access "abs(" and copy it to the home screen, we could have pressed $\boxed{\text{MATH}}$ $\boxed{\triangleright}$ 1 since "abs(" is item 1 on the MATH NUM menu. Note that, in general, an item can be selected from a menu in two ways. Either press the number or letter shown at the left of the item or use the $\boxed{\triangledown}$ or $\boxed{\triangle}$ key to highlight the item and then press $\boxed{\text{ENTER}}$.

Absolute value notation can also be found as the first item in the CATALOG and copied to the home screen. To do this press $\boxed{\text{2nd}}$ $\boxed{\text{CATALOG}}$ $\boxed{\text{ENTER}}$. (CATALOG is the second operation associated with the 0 numeric key.)

EDITING ENTRIES

After you have performed a computation, you can recall and edit an entry if necessary. Suppose, for instance, in entering one of the expressions in Example 3 above you pressed 6 instead of 3. To correct this, first press 2nd ENTRY to return to the last entry. (ENTRY is the second operation associated with the ENTER key.) Then use the ◁ key to move the cursor to 6 and press 3 to overwrite it. If you forgot to type the 2 in the first expression, move the cursor to the subtraction symbol; then press 2nd INS 2 to insert the 2 before that symbol. (INS is the second operation associated with the DEL key.) You can continue to insert symbols immediately after the first insertion without pressing 2nd INS again. If you typed 21 instead of 2, move the cursor to 1 and press DEL. This will delete the 1. If you notice that an entry needs to be edited before you press ENTER to perform the computation, the editing can be done directly without recalling the entry.

The keystrokes 2nd ENTRY can be used repeatedly to recall entries preceding the last one. Pressing 2nd ENTRY twice, for example, will recall the next to last entry. Using these keystrokes a third time recalls the third to last entry and so on. The number of entries that can be recalled depends on the amount of storage they occupy in the calculator's memory.

SCIENTIFIC NOTATION

To enter a number in scientific notation, first type the decimal portion of the number; then press 2nd EE (EE is the second operation associated with the , key.); finally type the exponent, which can be at most two digits. For example, to enter 1.789×10^{-11} in scientific notation, press 1 . 7 8 9 2nd EE (−) 1 1 ENTER. To enter 6.084×10^{23} in scientific notation, press 6 . 0 8 4 2nd EE 2 3 ENTER. The decimal portion of each number appears before a small E while the exponent follows the E.

```
1.789E -11
            1.789E -11
6.084E23
             6.084E23
```

The graphing calculator can be used to perform computations in scientific notation.

Section R.2, Example 8 *Distance to a Star.* The nearest star, Alpha Centauri C, is about 4.22 light-years from Earth. One light-year is the distance that light travels in one year and is about 5.88×10^{12} miles. How many miles is it from Earth to Alpha Centauri C? Express your answer in scientific notation.

To solve this problem we find the product $4.22 \times (5.88 \times 10^{12})$. Press 4 $\boxed{.}$ 2 2 $\boxed{\times}$ 5 $\boxed{.}$ 8 8 $\boxed{\text{2nd}}$ $\boxed{\text{EE}}$ 1 2 $\boxed{\text{ENTER}}$. The result is 2.48136×10^{13} miles.

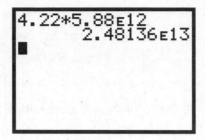

ORDER OF OPERATIONS

Section R.2, Example 9 (b) Calculate: $\dfrac{10 \div (8 - 6) + 9 \cdot 4}{2^5 + 3^2}$.

In order to divide the entire numerator by the entire denominator, we must enclose both the numerator and the denominator in parentheses. That is, we enter $(10 \div (8 - 6) + 9 \cdot 4) \div (2^5 + 3^2)$. Press $\boxed{(}$ 1 0 $\boxed{\div}$ $\boxed{(}$ 8 $\boxed{-}$ 6 $\boxed{)}$ $\boxed{+}$ 9 $\boxed{\times}$ 4 $\boxed{)}$ $\boxed{\div}$ $\boxed{(}$ 2 $\boxed{\wedge}$ 5 $\boxed{+}$ 3 $\boxed{x^2}$ $\boxed{)}$ $\boxed{\text{ENTER}}$. Note that 3^2 can be entered either as 3 $\boxed{x^2}$ or as 3 $\boxed{\wedge}$ 2.

```
(10/(8-6)+9*4)/(
2^5+3²)
                1
```

THE TVM SOLVER

Section R.2, Example 10 *Compound Interest.* If a principal P is invested at an interest rate r, compounded n times per year, in t years it will grow to an amount A given by

$$A = P\left(1 + \frac{r}{n}\right)^{nt}.$$

Suppose that \$1250 is invested at 4.6% interest, compounded quarterly. How much is in the account at the end of 8 yr?

To find the desired value, we could use a calculator to do the computation shown on page 14 of the text or we could use the calculator's TVM Solver. Here we will illustrate the use of the TVM Solver. This application is accessed by pressing $\boxed{\text{APPS}}$ $\boxed{\text{ENTER}}$ or $\boxed{\text{APPS}}$ 1 to select the Finance application and then press $\boxed{\text{ENTER}}$ or 1 to select the TVM Solver. Enter $4 \cdot 8$, or 32 (the total number of compounding periods) for N. Then use the $\boxed{\triangledown}$ key to position the cursor beside I% and enter 4.6 (the interest rate). Continuing in the same manner, enter PV = −1250 (the amount invested, entered as a negative number), PMT = 0, P/Y = 4, and C/Y = 4 (the number of compounding periods each year). The END/BEGIN setting at the bottom of the screen is irrelevant in this situation. Position the cursor beside FV = and press ALPHA SOLVE to display the desired value. (SOLVE is the Alpha operation associated with the ENTER key.) We see that there will be $1802.26 in the account at the end of 8 yr.

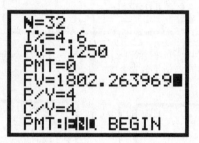

RADICAL NOTATION

We can use the square-root, cube-root, and xth-root features to simplify radical expressions.

Section R.7, Example 1 Simplify each of the following.

a) $\sqrt{36}$ b) $-\sqrt{36}$ c) $\sqrt[3]{-8}$ d) $\sqrt[5]{\dfrac{32}{243}}$ e) $\sqrt[4]{-16}$

a) To find $\sqrt{36}$, press 2nd $\sqrt{\ }$ 3 6) ENTER. ($\sqrt{\ }$ is the second operation associated with the x^2 key.) Note that the calculator supplies a left parenthesis with the radical symbol and we close the expression with a right parenthesis.

b) To find $-\sqrt{36}$, press (−) 2nd $\sqrt{\ }$ 3 6) ENTER.

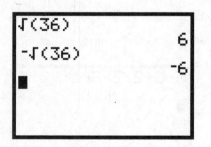

c) We will use the cube-root feature from the MATH menu to find $\sqrt[3]{-8}$. Press MATH 4 (−) 8) ENTER.

d) We will use the xth-root feature from the MATH menu to find $\sqrt[5]{\dfrac{32}{243}}$. We will also use ▷Frac to express the result as a fraction. Press 5 MATH 5 (3 2 ÷ 2 4 3) MATH 1 ENTER. The first 5 is the index of the radical, and the second 5 is used to select item 5, the xth-root, from the MATH menu.

e) To enter $\sqrt[4]{-16}$ press 4 $\boxed{\text{MATH}}$ 5 $\boxed{(-)}$ 1 6 $\boxed{\text{ENTER}}$. When the calculator is set in REAL mode, we get an error message indicating that the answer is nonreal.

```
4×√-16█
```

```
ERR:NONREAL ANS
1█Quit
2:Goto
```

RATIONAL EXPONENTS

We can add and subtract rational exponents on a graphing calculator.

Section R.7, Example 9 (a) Simplify and then, if appropriate, write radical notation for $x^{5/6} \cdot x^{2/3}$.

To find this product we first add the exponents, expressing the sum as a fraction. Press 5 $\boxed{\div}$ 6 $\boxed{+}$ 2 $\boxed{\div}$ 3 $\boxed{\text{MATH}}$ 1 $\boxed{\text{ENTER}}$. These keystrokes tell the calculator to add $\dfrac{5}{6}$ and $\dfrac{2}{3}$; then they access the MATH submenu of the MATH menu, copy item 1 "▷Frac" to the home screen, and display the result as a fraction.

```
5/6+2/3▶Frac
            3/2
█
```

Then, proceeding as on page 51 of the text, we find that the final result in radical notation is $x\sqrt{x}$.

Chapter 1
Graphs, Functions, and Models

GRAPHING EQUATIONS

Section 1.1, Example 4 Graph $3x - 5y = -10$.

First we solve for y as shown in the text, obtaining $y = \dfrac{3}{5}x + 2$. Then press $\boxed{Y=}$ to access the equation-editor screen. Clear any equations that are present. To clear an entry for Y_1, for example, position the cursor beside " $Y_1 =$" and press $\boxed{\text{CLEAR}}$. Do this for each existing entry. Also turn off any plots that are turned on. If the name of a plot is highlighted at the top of the equation-editor screen, it is turned on. To turn it off, position the cursor over it and press $\boxed{\text{ENTER}}$. The name will no longer be highlighted, indicating that it is now turned off.

Next enter the equation by positioning the cursor beside "$Y_1 =$" and pressing $\boxed{(}$ 3 $\boxed{\div}$ 5 $\boxed{)}$ $\boxed{X, T, \Theta, n}$ $\boxed{+}$ 2. Although the parentheses are not necessary, the equation is more easily read when they are used.

```
Plot1  Plot2  Plot3
\Y1◼(3/5)X+2■
\Y2=
\Y3=
\Y4=
\Y5=
\Y6=
\Y7=
```

Next we set the viewing window. This is the portion of the coordinate plane that appears on the calculator's screen. It is defined by the minimum and maximum values of x and y: Xmin, Xmax, Ymin, and Ymax. The notation [Xmin, Xmax, Ymin, Ymax] is used to represent these window settings or dimensions. For example, [−12, 12, −8, 8] denotes a window that displays the portion of the x-axis from −12 to 12 and the portion of the y-axis from −8 to 8. In addition, the distance between tick marks on the axes is defined by the settings Xscl and Yscl. In this manual Xscl and Yscl will be assumed to be 1 unless noted otherwise. The Xres setting sets the pixel resolution. We usually select Xres = 1. The window corresponding to the settings [−20, 30, −12, 20], Xscl = 5, Yscl = 2, Xres = 1, is shown below.

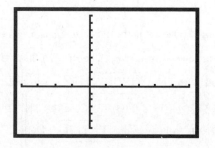

Press the $\boxed{\text{WINDOW}}$ key on the top row of the keypad to display the current window settings on your graphing calculator. The standard settings are shown below.

```
WINDOW
 Xmin=-10
 Xmax=10
 Xscl=1
 Ymin=-10
 Ymax=10
 Yscl=1
 Xres=1
```

To change a setting, position the cursor beside the setting you wish to change and enter the new value. For example, to change from the standard settings to $[-20,\ 30,\ -12,\ 20]$, Xscl = 5, Yscl = 2, on the WINDOW screen press $\boxed{(-)}$ 2 0 $\boxed{\text{ENTER}}$ 3 0 $\boxed{\text{ENTER}}$ 5 $\boxed{\text{ENTER}}$ $\boxed{(-)}$ 1 2 $\boxed{\text{ENTER}}$ 2 0 $\boxed{\text{ENTER}}$ 2 $\boxed{\text{ENTER}}$. Recall that the $\boxed{(-)}$ key in the bottom row of the keypad must be used to enter a negative number. The $\boxed{-}$ key in the right column of the keypad is used for the subtraction operation. The $\boxed{\triangledown}$ key may be used instead of $\boxed{\text{ENTER}}$ after typing each window setting. To see this window, press the $\boxed{\text{GRAPH}}$ key on the top row of the keypad. This is the window shown on the previous page.

QUICK TIP: To return quickly to the standard window setting $[-10,\ 10,\ -10,\ 10]$, Xscl = 1, Yscl = 1, press $\boxed{\text{ZOOM}}$ 6.

The standard window is a good choice for the graph of the equation $y = \dfrac{3}{5}x + 2$. Either enter these dimensions in the WINDOW screen and then press $\boxed{\text{GRAPH}}$ to see the graph or simply press $\boxed{\text{ZOOM}}$ 6 to select the standard window and see the graph.

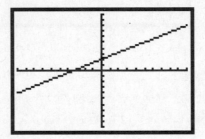

THE TABLE FEATURE

A table of x-and y-values representing ordered pairs that are solutions of an equation can be displayed on a graphing calculator.

Section 1.1, Example 5 Create a table of ordered pairs that are solutions of the equation $y = x^2 - 9x - 12$.

First we enter the equation on the equation-editor screen. Then press $\boxed{\text{2nd}}$ $\boxed{\text{TblSet}}$ to display the table set-up screen. (TblSet is the second function associated with the $\boxed{\text{WINDOW}}$ key.) You can choose to supply the x-values yourself or you can set the graphing calculator to supply them. To have the graphing calculator supply the x-values, set "Indpnt" to "Auto" by positioning the cursor over "Auto" and pressing $\boxed{\text{ENTER}}$. "Depend" should also be set to "Auto."

When "Indpnt" is set to "Auto," the graphing calculator will supply values for x, beginning with the value specified as TblStart and continuing by adding the value of ΔTbl to the preceding value for x. We will display a table of values that starts with $x = -3$ and adds 1 to the preceding x-value. Press $\boxed{(-)}$ 3 $\boxed{\triangledown}$ 1 or $\boxed{(-)}$ 3 $\boxed{\text{ENTER}}$ 1 to select a minimum x-value of -3 and an increment of 1. To display the table press $\boxed{\text{2nd}}$ $\boxed{\text{TABLE}}$. (TABLE is the second operation associated with the $\boxed{\text{GRAPH}}$ key.) We can use the $\boxed{\triangle}$ and $\boxed{\triangledown}$ keys to scroll up and down in the table to find values other than those shown here.

```
TABLE SETUP
 TblStart=-3
 ∆Tbl=1
Indpnt: AUTO  Ask
Depend: AUTO  Ask
```

```
  X   | Y1  |
 -3   | 24  |
 -2   | 10  |
 -1   | -2  |
  0   | -12 |
  1   | -20 |
  2   | -26 |
  3   | -30 |
X= -3
```

GRAPHING CIRCLES

If the center and radius of a circle are known, the circle can be graphed using the Circle feature from the DRAW menu.

Section 1.1, Example 12 Graph $(x-2)^2 + (y+1)^2 = 16$.

The center of this circle is $(2, -1)$ and its radius is 4. To graph it using the Circle feature from the DRAW menu first press $\boxed{\text{Y} =}$ and clear all previously entered equations. Then select a square window. (See pages 72 and 73 of the text for a discussion on squaring the viewing window.) We will use $[-9, 9, -6, 6]$. Press $\boxed{\text{2nd}}$ $\boxed{\text{QUIT}}$ to go to the home screen. Then press $\boxed{\text{2nd}}$ $\boxed{\text{DRAW}}$ 9 to display "Circle(." (DRAW is the second operation associated with the $\boxed{\text{PRGM}}$ key.) Enter the coordinates of the center and the radius, separating the entries by commas, and close the parentheses: 2 $\boxed{,}$ $\boxed{(-)}$ 1 $\boxed{,}$ 4 $\boxed{)}$ $\boxed{\text{ENTER}}$.

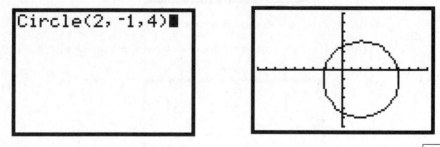

```
Circle(2,-1,4)■
```

To clear this circle from the Graph screen, use the ClrDraw option from the DRAW menu. Press $\boxed{\text{2nd}}$ $\boxed{\text{DRAW}}$ $\boxed{\text{ENTER}}$ or $\boxed{\text{2nd}}$ $\boxed{\text{DRAW}}$ 1 to do this.

FINDING FUNCTION VALUES

When a formula for a function is given, function values can be found in several ways.

Section 1.2, Example 4 (b) For $f(x) = 2x^2 - x + 3$, find $f(-7)$.

Method 1: Substitute the inputs directly in the formula. Press 2 $\boxed{(}$ $\boxed{(-)}$ 7 $\boxed{)}$ $\boxed{x^2}$ $\boxed{-}$ $\boxed{(}$ $\boxed{(-)}$ 7 $\boxed{)}$ $\boxed{+}$ 3 $\boxed{\text{ENTER}}$. Although it is not necessary to use the second set of parentheses, they allow the expression to be read more easily so we include them here.

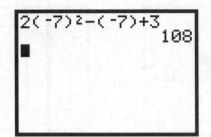

Method 2: Enter $y_1 = 2x^2 - x + 3$ on the "Y =" screen. Then press $\boxed{\text{2nd}}$ $\boxed{\text{QUIT}}$ to go to the home screen. To find $f(-7)$, the value of y_1 when $x = -7$, press $\boxed{(-)}$ 7 $\boxed{\text{STO} \triangleright}$ $\boxed{\text{X, T, } \Theta, \, n}$ $\boxed{\text{ALPHA}}$ $\boxed{:}$ $\boxed{\text{VARS}}$ $\boxed{\triangleright}$ 1 1 $\boxed{\text{ENTER}}$. (: is the ALPHA operation associated with the $\boxed{.}$ key.) This series of keystrokes stores -7 as the value of x and then substitutes it in the function y_1.

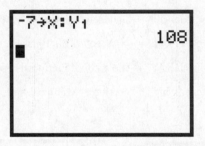

Method 3: Enter $y_1 = 2x^2 - x + 3$ on the "Y =" screen and press $\boxed{\text{2nd}}$ $\boxed{\text{QUIT}}$ to go to the home screen. To find $f(-7)$, press $\boxed{\text{VARS}}$ $\boxed{\triangleright}$ 1 1 $\boxed{(}$ $\boxed{(-)}$ 7 $\boxed{)}$ $\boxed{\text{ENTER}}$. Note that this entry closely resembles function notation.

Method 4: The TABLE feature can also be used to find function values. Enter $y_1 = 2x^2 - x + 3$ on the "Y =" screen. Then set up a table in Ask mode, by pressing $\boxed{\text{2nd}}$ $\boxed{\text{TBLSET}}$, moving the cursor over "Indpnt: Ask," and pressing $\boxed{\text{ENTER}}$. In Ask mode, you supply x-values and the calculator returns the corresponding y-values. The settings for TblStart and

ΔTbl are irrelevant in this mode. Press $\boxed{\text{2nd}}$ $\boxed{\text{TABLE}}$ to display the TABLE screen. Then press $\boxed{(-)}$ 7 $\boxed{\text{ENTER}}$ to find $f(-7)$.

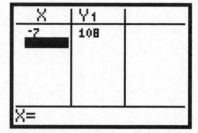

Method 5: We can also use the Value feature from the CALC menu to find $f(-7)$. To do this, graph $y_1 = 2x^2 - x + 3$ in a window that includes the x-value -7. We will use the standard window. Then press $\boxed{\text{2nd}}$ $\boxed{\text{CALC}}$ 1 or $\boxed{\text{2nd}}$ $\boxed{\text{CALC}}$ $\boxed{\text{ENTER}}$ to access the CALC menu and select item 1, Value. Now supply the desired x-value by pressing $\boxed{(-)}$ 7. Press $\boxed{\text{ENTER}}$ to see X $= -7$, Y $= 108$ at the bottom of the screen, Thus, $f(-7) = 108$.

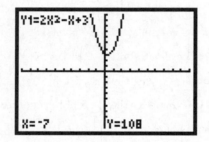

GRAPHS OF FUNCTIONS

Three functions are graphed in **Section 1.2, Example 5**. The TI-83 Plus and TI-84 Plus graphing calculators do not use function notation. Consequently, we must first replace function notation with y when we graph a function on one of these calculators. For example, to graph $f(x) = x^2 - 5$ replace $f(x)$ with y. Then enter the equation $y = x^2 - 5$ on the equation-editor screen and graph it as described on page 9 of this manual.

LINEAR REGRESSION

We can use the Linear Regression feature in the STAT CALC menu to fit a linear equation to a set of data.

Section 1.4, Example 7 The following table shows the number of U. S. households subscribing to cable television, in millions, for the years 1999 through 2006. Fit a regression line to the data using the linear regression feature on a graphing calculator. Then use the linear model to estimate the number of cable television subscribers in 2010.

Years, x	U. S. Households with Cable Television (in millions)
1999, 0	76.4
2000, 1	78.6
2001, 2	81.5
2002, 3	87.8
2003, 4	88.4
2004, 5	92.4
2005, 6	94.0
2006, 7	95.0

We will enter the data as ordered pairs on the STAT list editor screen. To clear any existing lists press $\boxed{\text{STAT}}$ 4 $\boxed{\text{2nd}}$ $\boxed{\text{L}_1}$, $\boxed{\text{2nd}}$ $\boxed{\text{L}_2}$, $\boxed{\text{2nd}}$ $\boxed{\text{L}_3}$, $\boxed{\text{2nd}}$ $\boxed{\text{L}_4}$, $\boxed{\text{2nd}}$ $\boxed{\text{L}_5}$, $\boxed{\text{2nd}}$ $\boxed{\text{L}_6}$ $\boxed{\text{ENTER}}$. (L_1 through L_6 are the second operations associated with the numeric keys 1 through 6.) The lists can also be cleared by first accessing the STAT list editor screen by pressing $\boxed{\text{STAT}}$ $\boxed{\text{ENTER}}$ or $\boxed{\text{STAT}}$ 1. These keystrokes display the STAT EDIT menu and then select the Edit option from that menu. Then, for each list that contains entries, use the arrow keys to move the cursor to highlight the name of the list at the top of the column and then press $\boxed{\text{CLEAR}}$ $\boxed{\bigtriangledown}$ or $\boxed{\text{CLEAR}}$ $\boxed{\text{ENTER}}$.

Once the lists are cleared, we can enter the coordinates of the points. We will enter the first coordinates in L_1 as the number of years since 1999 and the second coordinates in L_2. Position the cursor at the top of column L_1, below the L_1 heading. To enter 0 press 0 $\boxed{\text{ENTER}}$. Continue typing the first coordinates, 1, 2, 3, 4, 5, 6, and 7 in order, each followed by $\boxed{\text{ENTER}}$. The entries can be followed by $\boxed{\bigtriangledown}$ rather than $\boxed{\text{ENTER}}$ if desired. Press $\boxed{\triangleright}$ to move to the top of column L_2. Type the second coordinates, 76.4, 78.6, 81.5, 87.8, 88.4, 92.4, 94.0, and 95.0, in succession, each followed by $\boxed{\text{ENTER}}$ or $\boxed{\bigtriangledown}$. Note that the coordinates of each point must be in the same position in both lists.

The graphing calculator can then plot these points, creating a scatterplot of the data. To do this we turn on the STAT PLOT feature. To access the STAT PLOT screen, press $\boxed{\text{2nd}}$ $\boxed{\text{STAT PLOT}}$. (STAT PLOT is the second operation associated with the $\boxed{\text{Y =}}$ key in the upper left-hand corner of the keypad.) We will use Plot 1. Access it by highlighting 1 and pressing $\boxed{\text{ENTER}}$ or simply by pressing 1. Position the cursor over On and press $\boxed{\text{ENTER}}$ to turn on Plot 1. Now we select Type, Xlist, Ylist, and Mark. Type determines the style of the plot. We will select the first type shown, a scatterplot. Xlist and Ylist designate the STAT lists that supply the first and second coordinates of the points in the plot. The last item, Mark, allows us to choose a box, a cross, or a dot for each point. Here we have selected a box. To select Type and Mark, position the cursor over the appropriate selection and press $\boxed{\text{ENTER}}$. Use the L_1 and L_2 keys (associated with the

1 and 2 numeric keys) to select L_1 and L_2 as Xlist and Ylist, respectively. The entries should be as shown below.

If the plot has previously been set up as desired, you do not need to follow the steps above. Instead, you can turn the plot on from the equation-editor, or "Y =", screen. To use this alternative, press $\boxed{Y =}$ to go to this screen. Then, assuming Plot 1 has not yet been turned on, position the cursor over Plot 1 and press $\boxed{\text{ENTER}}$. Plot 1 will now be highlighted.

Note that there should be no equations entered on the equation-editor screen. If there are equations present, clear them as described on page 9 of this manual. If this is not done, the equations that are currently entered will be graphed along with the data points that are entered.

Now we select a window. The x-values range from 0 through 7 and the y-values from 76.4 through 95.0, so one good choice is $[-1, 8, 70, 100]$, Xscl =1, Yscl =5. We can enter these dimensions and then press $\boxed{\text{GRAPH}}$ to see the scatterplot. Instead of entering the window dimensions directly, we can press $\boxed{\text{ZOOM}}$ 9 after entering the coordinates of the points in lists, turning on Plot 1, and selecting Type, Xlist, Ylist, and Mark. This activates the ZoomStat operation which automatically defines a viewing window that displays all the points. We use the window $[-1, 8, 70, 100]$, Xscl =1, Yscl =5 to produce the scatterplot shown below.

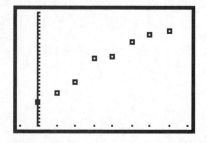

The calculator's linear regression feature can be used to fit a linear equation to the data. Once the data have been entered in the lists, press $\boxed{\text{STAT}}$ $\boxed{\triangleright}$ 4 $\boxed{\text{ENTER}}$ to select LinReg($ax + b$) from the STAT CALC menu and to display the coefficients a and b of the regression equation $y = ax + b$.

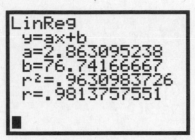

If the diagnostics have been turned on, values for r^2 and r will be displayed. These numbers indicate how well the regression line fits the data. See the discussion of the coefficient of linear correlation, r, on page 123 of the text.

If you wish to select DiagnosticOn mode, press ⎣2nd⎦ ⎣CATALOG⎦ and use ⎣▽⎦ to position the triangular selection cursor beside DiagnosticOn. To alleviate the tedium of scrolling through many items to reach DiagnosticOn, press ⎣D⎦ after pressing ⎣2nd⎦ ⎣CATALOG⎦ to move quickly to the first catalog item that begins with the letter D. (D is the ALPHA operation associated with the ⎣x^{-1}⎦ key.) Note that it is not necessary to press ⎣ALPHA⎦ before ⎣D⎦ when the catalog is displayed. Then use ⎣▽⎦ to scroll to DiagnosticOn. Press ⎣ENTER⎦ to paste this instruction to the home screen and then press ⎣ENTER⎦ a second time to set the mode. To select DiagnosticOff mode, press ⎣2nd⎦ ⎣CATALOG⎦, position the selection cursor beside DiagnosticOff, press ⎣ENTER⎦ to paste this instruction to the home screen, and then press ⎣ENTER⎦ again to set this mode.

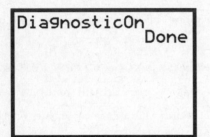

Immediately after the regression equation is found it can be copied to the equation-editor screen. (If you Selected DiagnosticOn *after* finding the regression equation, you will have to find the equation again in order to copy it to the equation-editor screen.) To copy the regression equation as Y_1, for example, first note that any previous entry in Y_1 should be cleared. Press ⎣Y =⎦ and position the cursor beside Y_1. Then press ⎣VARS⎦ ⎣5⎦ ⎣▷⎦ ⎣▷⎦ 1. These keystrokes select Statistics from the VARS menu, then select the EQ (Equation) submenu, and finally select the RegEq (Regression Equation) from this submenu.

Before the regression equation is found, it is possible to select a y-variable to which it will be stored on the equation editor screen. After the data have been stored in the lists and the equation previously entered as Y_1 has been cleared, press STAT ▷ 4 VARS ▷ 1 1 ENTER . The coefficients of the regression equation will be displayed on the home screen, and the regression equation will also be stored as Y_1 on the equation-editor screen.

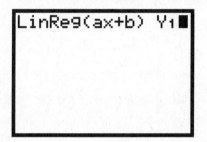

Once the regression equation has been copied to the equation-editor screen, we can see the graph of the equation along with the scatterplot by pressing GRAPH .

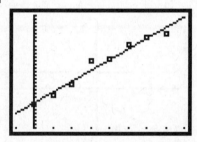

To estimate the number of U. S. households subscribing to cable television in 2010, evaluate the regression equation for $x = 11$. (2010 is 11 years after 1999.) Use any of the methods for evaluating a function presented earlier in this chapter. (See pages 12 and 13 of this manual.) We will use function notation on the home screen. When $x = 11, y \approx 108.2$, so we estimate that there will be about 108.2 million U. S. households subscribing to cable television in 2010.

Turn off the STAT PLOT as described on page 9 of this manual before graphing other functions.

THE INTERSECT METHOD

We can use the Intersect feature from the CALC menu to solve equations. We call this the **Intersect method**.

Section 1.5, Example 1 Solve $\frac{3}{4}x - 1 = \frac{7}{5}$.

Press $\boxed{\text{Y}=}$ to go to the equation-editor screen. Clear any existing entries and then enter $y_1 = \frac{3}{4}x - 1$ and $y_2 = \frac{7}{5}$. The solution of the original equation is the first coordinate of the point of intersection of the graphs of y_1 and y_2. Graph the equations in a window that displays the point of intersection. The window $[-5, 5, -5, 5]$ is a good choice.

Next we press $\boxed{\text{2nd}}$ $\boxed{\text{CALC}}$ 5 to select the Intersect feature from the CALC menu. (CALC is the second operation associated with the $\boxed{\text{TRACE}}$ key.) The query "First curve?" appears at the bottom of the screen. The blinking cursor is positioned on the graph of y_1. This is indicated by the notation $Y_1 = (3/4)X - 1$ in the upper left-hand corner of the screen. Press $\boxed{\text{ENTER}}$ to indicate that this is the first curve involved in the intersection. Next the query "Second curve?" appears at the bottom of the screen. The blinking cursor is now positioned on the graph of y_2 and the notation $Y_2 = 7/5$ should appear in the top left-hand corner of the screen. Press $\boxed{\text{ENTER}}$ to indicate that this is the second curve. We identify the curves for the calculator since we could have as many as ten graphs on the screen at once. After we identify the second curve, the query "Guess?" appears at the bottom of the screen. Use the right and left arrow keys to move the blinking cursor close to the point of intersection of the graphs or key in a number that appears to be close to the x-coordinate of this point. This provides the calculator with a guess as to the coordinates of this point. We do this since some pairs of curves can have more than one point of intersection. When the cursor is positioned or the x-value is entered, press $\boxed{\text{ENTER}}$ a third time. Now the coordinates of the point of intersection appear at the bottom of the screen.

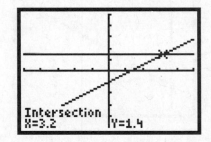

We see that the first coordinate of the point of intersection is 3.2. We can find fraction notation for the solution by using the ▷Frac feature from the MATH menu. To do this first press $\boxed{\text{2nd}}$ $\boxed{\text{QUIT}}$ to go to the home screen. The x- and y-coordinates of the point of intersection are stored in the calculator as X and Y, respectively. To convert the decimal notation for X to a fraction, press $\boxed{\text{X, T, }\Theta, n}$ $\boxed{\text{MATH}}$ 1 $\boxed{\text{ENTER}}$. These keystrokes tell the calculator to use X, and then they access the MATH submenu of the MATH menu, copy item 1 "▷ Frac" to the home screen, and display the conversion. We see that the first coordinate of the point of intersection is $\frac{16}{5}$. The solution of the equation is 3.2, or $\frac{16}{5}$.

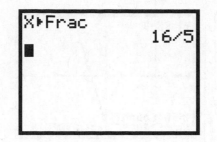

THE ZERO METHOD

The Zero feature from the CALC menu can be used to find the zeros of a function or to solve an equation in the form $f(x) = 0$. We call this the **Zero method**.

Section 1.5, Example 11 Find the zero of $f(x) = 5x - 9$.

On the equation-editor screen, clear any existing entries and then enter $y_1 = 5x - 9$. Now graph the function in a viewing window that shows the x-intercept clearly. The standard window is a good choice.

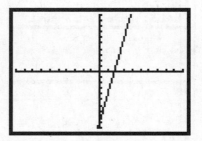

Press $\boxed{\text{2nd}}$ $\boxed{\text{CALC}}$ to display the CALC menu. Then press 2 to select the Zero feature. We are first prompted to select a left bound for the zero. This means that we must choose an x-value that is to the left of the x-intercept. This can be done by using the left- and right-arrow keys to move the cursor to a point on the curve to the left of the x-intercept or by keying in a value less than the first coordinate of the intercept.

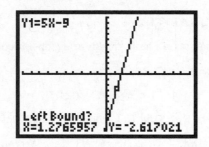

Once this is done press $\boxed{\text{ENTER}}$. Now we are prompted to select a right bound that is to the right of the x-intercept. Again, this can be done by using the arrow keys to move the cursor to a point on the curve to the right of the x-intercept or by keying in a value greater than the first coordinate of the intercept.

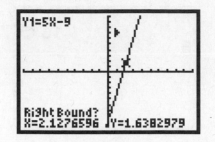

Press ENTER again. Finally we are prompted to make a guess as to the value of the zero. Move the cursor to a point close to the zero or key in a value.

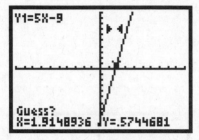

Finally, press ENTER a third time. We see that $y = 0$ when $x = 1.8$, so 1.8 is the zero of the function.

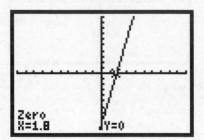

If a function has more than one zero, the Zero feature can be used as many times as necessary to find all of them.

CHECKING SOLUTIONS OF INEQUALITIES

We can perform a partial check of the solution of an inequality using operations from the TEST menu.

Section 1.6, Example 2 Solve: $-3 < 2x + 5 \leq 7$.

The solution set is found in the text. It is $\{x | -4 < x \leq 1\}$, or $(-4, 1]$. We can perform a partial check of this solution by graphing $y = (-3 < 2x + 5)$ *and* $(2x + 5 \leq 7)$ in Dot mode. To select Dot mode press MODE and use ▽ and ▷ to position the blinking cursor over "Dot." Then press ENTER. We can also select DOT mode by selecting the "dot" GraphStyle on the equation-editor screen. Position the cursor over the GraphStyle icon to the left of the equation to be graphed in Dot mode, and press ENTER repeatedly until the dotted icon appears. If the "line" icon was previously selected, ENTER must be pressed six times to select the "dot" style.

The value of y will be 1 for those x-values which make y a true statement. It will be 0 for those x-values for which y is false. To enter the expression for y, position the cursor beside Y_1 on the Y = screen. Then press $($ $(-)$ 3 2nd $\boxed{\text{TEST}}$ 5 2 $\boxed{\text{X, T, } \Theta, n}$ $+$ 5 $)$ 2nd $\boxed{\text{TEST}}$ $\triangleright$ 1 $($ 2 $\boxed{\text{X, T, } \Theta, n}$ $+$ 5 2nd $\boxed{\text{TEST}}$ 6 7 $)$. (TEST is the second operation associated with the $\boxed{\text{MATH}}$ key.) The keystrokes 2nd $\boxed{\text{TEST}}$ 5 and 2nd $\boxed{\text{TEST}}$ 6 display the TEST submenu of the TEST menu and paste the symbols "$<$" and "$\leq$," respectively, from that menu to the equation-editor screen. The keystrokes 2nd $\boxed{\text{TEST}}$ $\triangleright$ 1 display the LOGIC submenu of the TEST menu and paste "and" from that menu to the equation-editor screen.

Now select a window and press $\boxed{\text{GRAPH}}$. We use the window $[-10, 10, -1, 2]$.

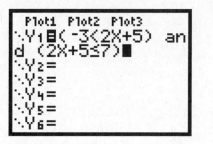 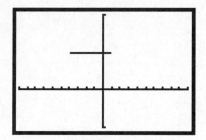

We see that $y = 1$ for x-values from -4 to 1, confirming that all x-values from -4 to 1 are in the solution set. The algebraic solution indicates that the endpoint 1 is also in the solution set.

Chapter 2
More on Functions

THE MAXIMUM AND MINIMUM FEATURES

Section 2.1, Example 2 Use a graphing calculator to determine any relative maxima or minima of the function $f(x) = 0.1x^3 - 0.6x^2 - 0.1x + 2$.

First graph $y_1 = 0.1x^3 - 0.6x^2 - 0.1x + 2$ in a window that displays the relative extrema of the function. Trial and error reveals that one good choice is $[-4, 6, -3, 3]$. Observe that a relative maximum occurs near $x = 0$ and a relative minimum occurs near $x = 4$.

To find the relative maximum, first press 2nd CALC 4 or 2nd CALC ▽ ▽ ▽ ENTER to select the Maximum feature from the CALC menu. We are prompted to select a left bound for the relative maximum. This means that we must choose an x-value that is to the left of the x-value of the point where the relative maximum occurs. This can be done by using the left- and right-arrow keys to move the cursor to a point to the left of the relative maximum or by keying in an appropriate value.

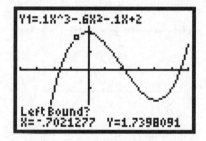

Once this is done, press ENTER . Now we are prompted to select a right bound. We move the cursor to a point to the right of the relative maximum or we can key in an appropriate value.

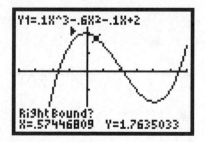

Press ENTER again. Finally we are prompted to guess the x-value at which the relative maximum occurs. Move the cursor close to the relative maximum point or key in an x-value.

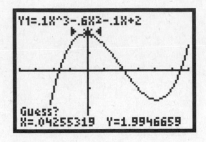

Press ENTER a third time. We see that a relative maximum function value of approximately 2.004 occurs when $x \approx -0.082$.

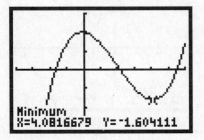

To find the relative minimum, select the Minimum feature from the CALC menu by pressing 2nd CALC 3 or 2nd CALC ▽ ▽ ENTER. Select left and right bounds for the relative minimum and guess the x-value at which it occurs as described above. We see that a relative minimum function value of approximately -1.604 occurs when $x \approx 4.082$.

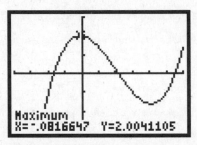

THE GREATEST INTEGER FUNCTION

The greatest integer function is found in the MATH NUM menu and is denoted "int." To find int(1.9) press MATH ▷ 5 1 . 9) ENTER. Note that the calculator supplies a left parenthesis and we close the parentheses after entering 1.9.

We can also graph the greatest integer function.

Section 2.1, Example 9 Graph $f(x) = \text{int}(x)$.

First we set the calculator in Dot mode. If the calculator is left in Connected mode, it will connect adjacent points, or pixels, with line segments and produce a graph of the greatest integer function that looks like stair-steps rather than a set of horizontal line segments. (See the procedure for selecting Dot mode on page 20 of this manual.)

With the calculator set in Dot mode, clear any functions that have previously been entered on the equation-editor screen. Then position the cursor beside "Y1 =" and select the greatest integer function from the MATH NUM menu by pressing $\boxed{\text{MATH}}$ $\boxed{\triangleright}$ 5 $\boxed{\text{X, T, }\Theta\text{, }n}$ $\boxed{)}$. Select a window and press $\boxed{\text{GRAPH}}$. The window $[-6, 6, -6, 6]$ is shown here.

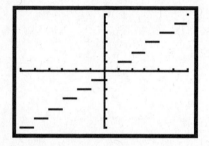

THE COMPOSITION OF FUNCTIONS

We can evaluate composite functions on a graphing calculator.

Section 2.3, Example 1 (b) Given that $f(x) = 2x - 5$ and $g(x) = x^2 - 3x + 8$, find $(f \circ g)(7)$ and $(g \circ f)(7)$.

On the equation-editor screen enter $y_1 = 2x - 5$ and $y_2 = x^2 - 3x + 8$. Then $(f \circ g)(7) = (y_1 \circ y_2)(7)$, or $y_1(y_2(7))$ and $(g \circ f)(7) = (y_2 \circ y_1)(7)$, or $y_2(y_1(7))$. To find these function values press $\boxed{\text{2nd}}$ $\boxed{\text{QUIT}}$ to go to the home screen. Then enter $y_1(y_2(7))$ by pressing $\boxed{\text{VARS}}$ $\boxed{\triangleright}$ 1 1 $\boxed{(}$ $\boxed{\text{VARS}}$ $\boxed{\triangleright}$ 1 2 $\boxed{(}$ 7 $\boxed{)}$ $\boxed{)}$ $\boxed{\text{ENTER}}$. Enter $y_2(y_1(7))$ by pressing $\boxed{\text{VARS}}$ $\boxed{\triangleright}$ 1 2 $\boxed{(}$ $\boxed{\text{VARS}}$ $\boxed{\triangleright}$ 1 1 $\boxed{(}$ 7 $\boxed{)}$ $\boxed{)}$ $\boxed{\text{ENTER}}$.

```
Y1(Y2(7))
                67
Y2(Y1(7))
                62
■
```

Chapter 3
Quadratic Functions and Equations; Inequalities

OPERATIONS WITH COMPLEX NUMBERS

Operations with complex numbers can be performed on the TI-83 Plus and the TI-84 Plus. First set the calculator in the complex $a + bi$ mode by pressing MODE , positioning the cursor over $a + bi$, and pressing ENTER . The TI-83 Plus Mode screen is shown here.

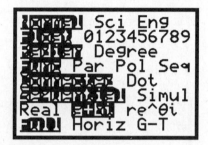

Section 3.1, Example 2

(a) Add: $(8 + 6i) + (3 + 2i)$.

To find this sum go to the home screen and press 8 + 6 2nd i + 3 + 2 2nd i ENTER . (The number i is the second operation associated with the · key.) Note that it is not necessary to include parentheses when we are adding.

(b) Subtract: $(4 + 5i) - (6 - 3i)$.

Press 4 + 5 2nd i − (6 − 3 2nd i) ENTER . Note that the parentheses must be included as shown so that the entire number $6 - 3i$ is subtracted.

```
8+6i+3+2i
            11+8i
4+5i-(6-3i)
            -2+8i
■
```

Section 3.1, Example 3

(a) Multiply: $\sqrt{-16} \cdot \sqrt{-25}$.

Press 2nd $\sqrt{}$ (−) 1 6) 2nd $\sqrt{}$ (−) 2 5) ENTER .

(b) Multiply: $(1 + 2i)(1 + 3i)$.

Press (1 + 2 2nd i) (1 + 3 2nd i) ENTER .

(c) Multiply: $(3 - 7i)^2$.

Press $\boxed{(}$ 3 $\boxed{-}$ 7 $\boxed{\text{2nd}}$ $\boxed{i}$ $\boxed{)}$ $\boxed{x^2}$ $\boxed{\text{ENTER}}$.

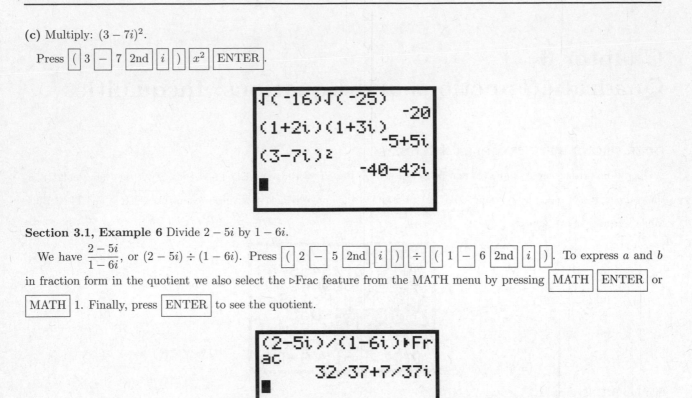

Section 3.1, Example 6 Divide $2 - 5i$ by $1 - 6i$.

We have $\dfrac{2 - 5i}{1 - 6i}$, or $(2 - 5i) \div (1 - 6i)$. Press $\boxed{(}$ 2 $\boxed{-}$ 5 $\boxed{\text{2nd}}$ $\boxed{i}$ $\boxed{)}$ $\boxed{\div}$ $\boxed{(}$ 1 $\boxed{-}$ 6 $\boxed{\text{2nd}}$ $\boxed{i}$ $\boxed{)}$. To express a and b in fraction form in the quotient we also select the ▷Frac feature from the MATH menu by pressing $\boxed{\text{MATH}}$ $\boxed{\text{ENTER}}$ or $\boxed{\text{MATH}}$ 1. Finally, press $\boxed{\text{ENTER}}$ to see the quotient.

Chapter 4
Polynomial and Rational Functions

POLYNOMIAL MODELS

We can fit polynomial functions to data on a graphing calculator.

Section 4.1, Example 9 The table below shows the number of miles of U. S.-owned operating railroad track for several years beginning with 1830. Model the data with a cubic function and with a quartic function. Let the first coordinate of each data point be the number of years after 1830.

Years, x	Miles of U. S.-Owned Operating Railroad Track
1830, 0	23
1840, 10	2,818
1850, 20	9,021
1860, 30	30,635
1870, 40	52,922
1880, 50	92,147
1890, 60	163,597
1900, 70	193,346
1910, 80	240,293
1916, 86	254,037
1920, 90	252,845
1930, 100	249,052
1940, 110	233,670
1950, 120	223,779
1960, 130	217,552
1970, 140	205,782
1980, 150	178,056
1990, 160	145,979
2000, 170	144,473
2003, 173	141,509
2005, 175	140,810

Enter the data in lists as described on page 14 of this manual. To model the data with a cubic function select cubic regression from the STAT CALC menu by pressing $\boxed{\text{STAT}}$ $\boxed{\triangleright}$ 6 $\boxed{\text{ENTER}}$. The graphing calculator displays the coefficients of a cubic function $y = ax^3 + bx^2 + cx + d$. The function is shown on the next page.

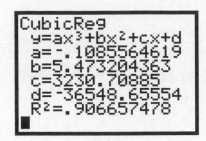

To model the data with a quartic function select quartic regression from the STAT CALC menu by pressing $\boxed{\text{STAT}}$ $\boxed{\triangleright}$ 7 $\boxed{\text{ENTER}}$. The graphing calculator displays the coefficients of a quartic function $y = ax^4 + bx^3 + cx^2 + dx + e$.

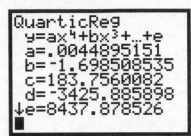   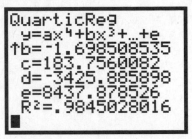

A scatterplot of the data can be graphed as described on pages 14 and 15 of this manual. This function can be copied to the Y = screen using one of the methods described on pages 16 and 17 of this manual. Then it can be graphed along with the scatterplot. It can also be evaluated using one of the methods on pages 12 and 13.

GRAPHING RATIONAL FUNCTIONS

Section 4.5, Example 1 Consider $f(x) = \dfrac{1}{x-3}$. Find the domain and graph f.

In the text the domain is found to be $\{x | x \neq 3\}$, or $(-\infty, 3) \cup (3, \infty)$. Thus, there is not a point on the graph with an x-coordinate of 3. Graphing the function in Connected mode can lead to an incorrect graph in which a line connects the last point plotted to the left of $x = 3$ with the first point plotted to the right of $x = 3$. This line can be eliminated by using Dot mode as described on page 20 of this manual. Selecting a ZDecimal window from the ZOOM menu will also produce a graph in which this line does not appear in Connected mode. To do this, first enter $y = \dfrac{1}{x-3}$ on the Y = screen and then press $\boxed{\text{ZOOM}}$ 4. The resulting window dimensions and graph are shown below.

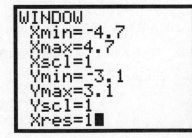   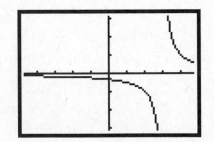

Section 4.5, Example 10 Graph: $g(x) = \dfrac{x-2}{x^2 - x - 2}$.

As explained in the text, the graph of $g(x)$ is the graph of $y = \dfrac{1}{x+1}$ with the point $\left(2, \dfrac{1}{3}\right)$ missing. To use a graphing calculator to produce the graph with a "hole" at $\left(2, \dfrac{1}{3}\right)$, use the ZDecimal feature from the ZOOM menu. After entering $y = \dfrac{x-2}{x^2 - x - 2}$ on the Y = screen, press $\boxed{\text{ZOOM}}$ 4 to select this window and display the graph. To confirm that there is no point on the graph with a first coordinate of 2, try to use the Value feature from the CALC menu to find the value of the function when $x = 2$. We see that there is not a y-value that corresponds to $x = 2$.

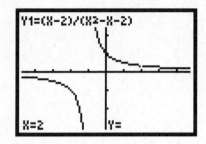

Chapter 5
Exponential and Logarithmic Functions

GRAPHING AN INVERSE FUNCTION

The DrawInv operation can be used to graph a function and its inverse on the same screen. A formula for the inverse function need not be found in order to do this. The calculator must be set in Func mode when this operation is used.

Section 5.1, Example 7 Graph $f(x) = 2x - 3$ and $f^{-1}(x)$ using the same set of axes.

Enter $y_1 = 2x - 3$, clear all other functions on the "Y =" screen, and select a window. We will use the standard window. (If you previously selected DOT mode from the MODE screen when graphing rational functions, return to that screen and select CONNECTED mode now.) Press $\boxed{\text{2nd}}$ $\boxed{\text{DRAW}}$ 8 to select the DrawInv operation. (DRAW is the second operation associated with the $\boxed{\text{PRGM}}$ key.) Follow these keystrokes with $\boxed{\text{VARS}}$ $\boxed{\triangleright}$ 1 1 to select function y_1. Press $\boxed{\text{ENTER}}$ to see the graph of the function and its inverse. The graphs are shown here in the standard window.

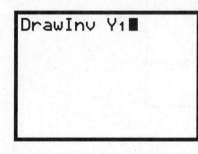
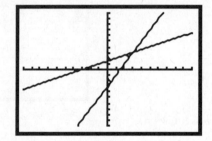

EVALUATING e^x, Log x, and Ln x

Use the calculator's scientific keys to evaluate e^x, $\log x$, and $\ln x$ for specific values of x.

Section 5.2, Example 5 (a), (b) Find the values of e^3 and $e^{-0.23}$. Round to four decimal places.

To find e^3 press $\boxed{\text{2nd}}$ $\boxed{e^x}$ 3 $\boxed{)}$ $\boxed{\text{ENTER}}$. (e^x is the second operation associated with the $\boxed{\text{LN}}$ key.) The calculator returns 20.08553692. Thus, $e^3 \approx 20.0855$. To find $e^{-0.23}$ press $\boxed{\text{2nd}}$ $\boxed{e^x}$ $\boxed{(-)}$ $\boxed{\cdot}$ 2 3 $\boxed{)}$ $\boxed{\text{ENTER}}$. The calculator returns .7945336025, so $e^{-0.23} \approx 0.7945$.

Section 5.3, Example 5 Find the values of log 645,778, log 0.0000239, and log (-3). Round to four decimal places.

To find log 645,778 press $\boxed{\text{LOG}}$ 6 4 5 7 7 8 $\boxed{)}$ $\boxed{\text{ENTER}}$ and read 5.810083246. Thus, log $645,778 \approx 5.8101$. To find log 0.0000239 press $\boxed{\text{LOG}}$ $\boxed{\cdot}$ 0 0 0 0 2 3 9 $\boxed{)}$ $\boxed{\text{ENTER}}$. The calculator returns -4.621602099, so log $0.0000239 \approx -4.6216$. When the calculator is set in Real mode the keystrokes $\boxed{\text{LOG}}$ $\boxed{(-)}$ 3 $\boxed{)}$ $\boxed{\text{ENTER}}$ produce the message ERR: NONREAL ANS, indicating that the result of this calculation is not a real number.

Section 5.3, Example 6 (a), (b), (c) Find the values of ln 645,778, ln 0.0000239, and ln (-5). Round to four decimal places.

To find ln 645,778 and ln 0.0000239 repeat the keystrokes used above to find log 645,778 and log 0.0000239 but press
$\boxed{\text{LN}}$ rather than $\boxed{\text{LOG}}$. We find that ln 645,778 ≈ 13.3782 and ln 0.0000239 ≈ −10.6416. When the calculator is set
in Real mode the keystrokes $\boxed{\text{LN}}$ $\boxed{(-)}$ $\boxed{5}$ $\boxed{)}$ $\boxed{\text{ENTER}}$ produce the message ERR: NONREAL ANS, indicating that the
result of this calculation is not a real number.

USING THE CHANGE OF BASE FORMULA

To find a logarithm with a base other than 10 or e we use the change-of-base formula, $\log_b M = \dfrac{\log_a M}{\log_a b}$, where a and
b are any logarithmic bases and M is any positive number.

Section 5.3, Example 7 Find $\log_5 8$ using common logarithms.

We let $a = 10$, $b = 5$, and $M = 8$ and substitute in the change-of-base formula. We have $\log_5 8 = \dfrac{\log_{10} 8}{\log_{10} 5}$. To perform
this computation press $\boxed{\text{LOG}}$ $\boxed{8}$ $\boxed{)}$ $\boxed{\div}$ $\boxed{\text{LOG}}$ $\boxed{5}$ $\boxed{)}$ $\boxed{\text{ENTER}}$. Note that the parentheses must be closed in the numerator
to enter the expression correctly. We also close the parentheses in the denominator for completeness. The result is about
1.2920. We could have let $a = e$ and used natural logarithms to find $\log_5 8$ as well, as shown in Example 8 in the text.

Section 5.3, Example 9 Graph $y = \log_5 x$.

To use a graphing calculator we must first change the base to e or 10. Here we use e. Let $a = e$, $b = 5$, and $M = x$
and substitute in the change-of-base formula. Enter $y_1 = \dfrac{\ln x}{\ln 5}$ on the Y = screen, select a window, and press $\boxed{\text{GRAPH}}$.
Note that, since the calculator forces the use of parentheses with the ln function, the parentheses in the numerator must
be closed: $\ln(x)/\ln(5)$. The right parenthesis following the 5 is optional but we include it for completeness.

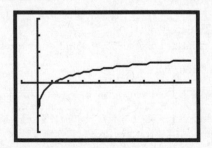

EXPONENTIAL AND LOGARITHMIC REGRESSION

In addition to the types of polynomial regression discussed earlier, exponential and logarithmic functions can be fit to
data. The operations of entering data, making scatterplots, and graphing and evaluating these functions are the same as

for linear regression functions. So are the procedures for copying a regression equation to the Y = screen, graphing it, and using it to find function values. Note that the coefficient of correlation, r, will be displayed only if DiagnosticOn has been selected from the Catalog.

Section 5.6, Example 6 *Surveillance Cameras.* The number of U. S. communities using surveillance cameras at intersections has increased greatly in recent years, as shown in the following table.

Year, x	Number of U. S. Communities Using Surveillance Cameras
1999, 0	19
2001, 2	35
2003, 4	75
2005, 6	130
2007, 8	243

(a) Use a graphing calculator to fit an exponential function to the data.

Enter the data in lists as described on page 14 of this manual. Then select exponential regression from the STAT CALC menu by pressing STAT ▷ 0 ENTER . The calculator displays the coefficient a and the base b for the exponential function $y = a \cdot b^x$.

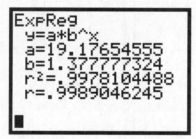

A scatterplot of the data can be graphed as described on pages 14 and 15 of this manual. This function can be copied to the Y = screen using one of the methods described on pages 16 and 17 of this manual. Then it can be graphed along with the scatterplot. It can also be evaluated using one of the methods on pages 12 and 13.

Section 5.6, Exercise 30 (a) *Forgetting.* In an art class, students were tested at the end of the course on a final exam. Then they were retested with an equivalent test at subsequent time intervals. Their scores after time x, in months, are given in the following table.

Time, x (in months)	Score, y
1	84.9%
2	84.6%
3	84.4%
4	84.2%
5	84.1%
6	83.9%

(a) Use a graphing calculator to fit a logarithmic function $y = a + b \ln x$ to the data.

After entering the data in lists as described on page 14 of this manual, press STAT ▷ to view the STAT CALC menu.

Select LnReg by pressing 9 $\boxed{\text{ENTER}}$. The values of a and b for the logarithmic function $y = a + b \ln x$ are displayed. This function can be evaluated using one of the methods described on pages 12 and 13 of this manual.

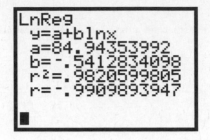

LOGISTIC REGRESSION

A logistic function can be fit to data using the TI-83, TI-83 Plus, or TI-84 Plus.

Section 5.6, Exercise 34 (a) *Effect of Advertising.* A company introduces a new software product on a trial run in a city. They advertised the product on television and found the following data relating the percent P of people who bought the product after x ads were run.

Number of Ads, x	Percent Who Bought, P
0	0.2
10	0.7
20	2.7
30	9.2
40	27
50	57.6
60	83.3
70	94.8
80	98.5
90	99.6

(a) Use a graphing calculator to fit a logistic function $P(x) = \dfrac{a}{1 + be^{-kx}}$ to the data.

After entering the data in lists as described on page 14 of this manual, press $\boxed{\text{STAT}}$ $\boxed{\triangleright}$ to view the STAT CALC menu. Select Logistic by pressing $\boxed{\text{ALPHA}}$ B $\boxed{\text{ENTER}}$. The values of a, b, and c for the logistic function $y = \dfrac{c}{1 + ae^{-bx}}$ are displayed. This function can be evaluated using one of the methods described on pages 12 and 13 of this manual.

```
Logistic
 y=c/(1+ae^(-bx)
 a=489.2438401
 b=.1299899024
 c=99.98884912
```

Chapter 6
The Trigonometric Functions

CONVERTING BETWEEN D°M′S″ AND DECIMAL DEGREE MEASURE

We can convert D°M′S″ notation to decimal notation and vice versa on the TI-83 Plus and the TI-84 Plus.

Section 6.1, Example 5 Convert 5°42′30″ to decimal degree notation.

Enter 5°42′30″ by pressing 5 $\boxed{\text{2nd}}$ $\boxed{\text{ANGLE}}$ 1 4 2 $\boxed{\text{2nd}}$ $\boxed{\text{ANGLE}}$ 2 3 0 $\boxed{\text{ALPHA}}$ $\boxed{''}$ $\boxed{\text{ENTER}}$. (ANGLE is the second operation associated with the $\boxed{\text{APPS}}$ key. ″ is the ALPHA operation associated with the $\boxed{+}$ key.) The calculator returns 5.708333333, so 5°42′30″ ≈ 5.71°.

Section 6.1, Exercise 39 Convert 15′5″ to decimal degree notation.

In converting from D°M′S″ notation to decimal degree notation, we must always enter a number for degrees even if it is 0. Thus, to convert 15′5″ to decimal degree notation, we enter 0°15′5″ as described above. The result is approximately 0.25°.

Section 6.1, Exercise 41 Convert 5°53″ to decimal degree notation.

All entries containing a nonzero number of seconds must include entries for the number of degrees and minutes even if one or both is 0. Thus, to convert 5°53″ to decimal degree notation, we enter 5°0′53″ as described in Example 5 above. The result is approximately 5.01°.

```
5°42'30"
          5.708333333
0°15'5"
           .2513888889
5°0'53"
          5.014722222
■
```

Section 6.1, Example 6 Convert 72.18°to D°M′S″ notation.

We use the ▷D°M′S″feature from the ANGLE menu to do this conversion. Press 7 2 $\boxed{\cdot}$ 1 8 $\boxed{\text{2nd}}$ $\boxed{\text{ANGLE}}$ 4 $\boxed{\text{ENTER}}$. The calculator returns 72°10′48″.

```
72.18▶DMS
            72°10'48"
■
```

FINDING TRIGONOMETRIC FUNCTION VALUES

A graphing calculator's SIN, COS, and TAN operations can be used to find the values of trigonometric functions.

Section 6.1, Example 7 Find the trigonometric function value, rounded to four decimal places, of each of the following.

a) tan 29.7°   b) sec 48°   c) sin 84°10′39″

a) With the calculator set in Degree mode, press ⸢TAN⸣ ⸢2⸣ ⸢9⸣ ⸢·⸣ ⸢7⸣ ⸢)⸣ ⸢ENTER⸣. Although it is not necessary to close the parentheses, we do it for completeness. If the calculator is set in Radian mode, press ⸢2nd⸣ ⸢ANGLE⸣ ⸢ENTER⸣ after the 7 to copy the degree symbol after the angle. This indicates to the calculator that the angle is given in degrees. We find that tan 29.7° ≈ 0.5704.

b) The secant, cosecant, and cotangent functions can be found by taking the reciprocals of the cosine, sine, and tangent functions, respectively. This can be done either by entering the reciprocal or by using the ⸢x^{-1}⸣ key. To find sec 48° we can enter the reciprocal of cos 48° on a TI-83 Plus or a TI-84 Plus set in Degree mode by pressing 1 ⸢÷⸣ ⸢COS⸣ ⸢4⸣ ⸢8⸣ ⸢)⸣ ⸢ENTER⸣. To find sec 48° using the ⸢x^{-1}⸣ key press ⸢COS⸣ ⸢4⸣ ⸢8⸣ ⸢)⸣ ⸢x^{-1}⸣ ⸢ENTER⸣. If the calculator is set in Radian mode, press ⸢2nd⸣ ⸢ANGLE⸣ ⸢ENTER⸣ after the 8 to copy the degree symbol after the angle. The result is sec 48° ≈ 1.4945. The screen below was produced with the calculator set in Degree mode.

c) With the calculator set in Degree mode, press $\boxed{\text{SIN}}$ followed by 84°10′39″ entered as described above in Converting Between D°M′S″ and Decimal Degree Measure. Then press $\boxed{\text{ENTER}}$. We find that sin 84°10′39″ ≈ 0.9948.

FINDING ANGLES

The inverse trigonometric function keys provide a quick way to find an angle given a trigonometric function value for that angle.

Section 6.1, Example 8 Find the acute angle, to the nearest tenth of a degree, whose sine value is approximately 0.20113.

Although the TABLE feature can be used to approximate this angle, it is faster to use the inverse sine key. With the calculator set in Degree mode, press $\boxed{\text{2nd}}$ $\boxed{\text{SIN}^{-1}}$ $\boxed{\cdot}$ 2 0 1 1 3 $\boxed{)}$ $\boxed{\text{ENTER}}$. (SIN^{-1} is the second operation associated with the $\boxed{\text{SIN}}$ key.) Although it is not necessary to close the parentheses in this case, we do it for completeness. We find that the desired acute angle is approximately 11.6°.

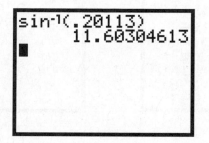

Section 6.1, Exercise 79 Find the acute angle, to the nearest tenth of a degree, whose cotangent value is 2.127.

Angles whose secant, cosecant, or cotangent values are known can be found using the reciprocals of the cosine, sine, and tangent functions, respectively. Since $\cot \theta = \dfrac{1}{\tan \theta} = 2.127$, we have $\tan \theta = \dfrac{1}{2.127}$, or $(2.127)^{-1}$. To find θ, press $\boxed{\text{2nd}}$ $\boxed{\text{TAN}^{-1}}$ 1 $\boxed{\div}$ 2 $\boxed{\cdot}$ 1 2 7 $\boxed{)}$ $\boxed{\text{ENTER}}$ or $\boxed{\text{2nd}}$ $\boxed{\text{TAN}^{-1}}$ 2 $\boxed{\cdot}$ 1 2 7 $\boxed{x^{-1}}$ $\boxed{)}$ $\boxed{\text{ENTER}}$. (TAN^{-1} is the second operation associated with the $\boxed{\text{TAN}}$ key.) Note that the parentheses are necessary in each set of keystrokes. The left parenthesis appears along with "$\tan^{-1}$" while the right parenthesis mst be keyed in. Without parentheses we would be finding the angle whose tangent is 1 and then dividing that angle by 2.127. We find that $\theta \approx 25.2°$.

```
tan-1(1/2.127)
         25.18036384
tan-1(2.127-1)
         25.18036384
■
```

CONVERTING BETWEEN DEGREE AND RADIAN MEASURE

We can use a graphing calculator to convert from degree to radian measure and vice versa. The calculator should be set in Radian mode when converting from degree to radian measure and in Degree mode when converting from radian to degree measure.

Section 6.4, Example 3 Convert each of the following to radians.

a) 120°   b) −297.25°

a) Set the calculator in Radian mode. Press 1 2 0 $\boxed{\text{2nd}}$ $\boxed{\text{ANGLE}}$ 1 $\boxed{\text{ENTER}}$ to enter 120°. The calculator returns a decimal approximation of the radian measure. We see that 120° ≈ 2.09 radians.

b) With the calculator set in Radian mode press $\boxed{(-)}$ 2 9 7 $\boxed{\cdot}$ 2 5 $\boxed{\text{2nd}}$ $\boxed{\text{ANGLE}}$ 1 $\boxed{\text{ENTER}}$. We see that −297.25° ≈ −5.19 radians.

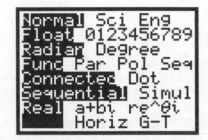

Section 6.4, Example 4 Convert each of the following to degrees.

a) $\dfrac{3\pi}{4}$ radians   b) 8.5 radians

a) Set the calculator in Degree mode. Then press $\boxed{(}$ 3 $\boxed{\text{2nd}}$ $\boxed{\pi}$ $\boxed{\div}$ 4 $\boxed{)}$ $\boxed{\text{2nd}}$ $\boxed{\text{ANGLE}}$ 3 $\boxed{\text{ENTER}}$ to enter $\dfrac{3\pi}{4}$ radians. (π is the second operation associated with the $\boxed{\wedge}$ key). The calculator returns 135, so $3\pi/4$ radians = 135°. Note that the parentheses are necessary in order to enter the entire expression in radian measure. Without the parentheses, the calculator reads only the denominator, 4, in radian measure and an incorrect result occurs.

b) With the calculator set in Degree mode press 8 $\boxed{\cdot}$ 5 $\boxed{\text{2nd}}$ $\boxed{\text{ANGLE}}$ 3 $\boxed{\text{ENTER}}$. The calculator returns 487.0141259, so 8.5 radians ≈ 487.01°.

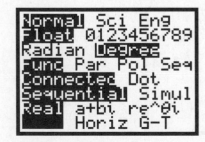
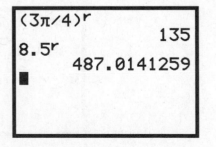

Chapter 7
Trigonometric Identities, Inverse Functions, and Equations

THE PATH GRAPH STYLE

Seven graph styles can be selected on the equation-editor screen of the TI-83 Plus and the TI-84 Plus. The path graph style can be used, along with the line style, to determine whether graphs coincide. This can be used to provide a partial check of an identity.

Section 7.1, Example 1 Use a graphing calculator to do a partial check of the identity $\cos x(\tan x - \sec x) = \sin x - 1$.

First press $\boxed{\text{MODE}}$ to determine whether Sequential mode is selected. If it is not, position the blinking cursor over Sequential and then press $\boxed{\text{ENTER}}$.

Next, on the Y = screen, enter $y_1 = \cos x(\tan x - 1/\cos x)$ and $y_2 = \sin x - 1$. Note that we entered $\sec x$ as $1/\cos x$. We will select the line graph style for y_1 and the path style for y_2. To select these graph styles use $\boxed{\triangleleft}$ to position the cursor over the icon to the left of the equation and press $\boxed{\text{ENTER}}$ repeatedly until the desired style icon appears as shown below.

The calculator will graph y_1 first as a solid line. Then y_2 will be graphed as the circular cursor traces the leading edge of the graph, allowing us to determine visually whether the graphs coincide. In this case, the graphs appear to coincide, so the identity is probably correct.

FINDING INVERSE FUNCTION VALUES

We can use a graphing calculator to find inverse function values in both radians and degrees.

Section 7.4, Example 2 (a), (e) Approximate $\cos^{-1}(-0.2689)$ and $\csc^{-1} 8.205$ in both radians and degrees.

To find inverse function values in radians, first set the calculator in Radian mode. Then, to approximate $\cos^{-1}(-0.2689)$, press $\boxed{\text{2nd}}$ $\boxed{\text{COS}^{-1}}$ $\boxed{(-)}$ $\boxed{\cdot}$ 2 6 8 9 $\boxed{)}$ $\boxed{\text{ENTER}}$. Although it is not necessary to close the parentheses in this case, we do it for completeness. The calculator returns 1.84304711, so $\cos^{-1}(-0.2689) \approx 1.8430$ radians.

To find $\csc^{-1} 8.205$, recall the identity $\csc\theta = \dfrac{1}{\sin\theta}$. Then $\csc^{-1} 8.205 = \sin^{-1}\left(\dfrac{1}{8.205}\right)$. Press $\boxed{\text{2nd}}$ $\boxed{\text{SIN}^{-1}}$ 1 $\boxed{\div}$ 8 $\boxed{\cdot}$ 2 0 5 $\boxed{)}$ $\boxed{\text{ENTER}}$ or $\boxed{\text{2nd}}$ $\boxed{\text{SIN}^{-1}}$ 8 $\boxed{\cdot}$ 2 0 5 $\boxed{x^{-1}}$ $\boxed{)}$ $\boxed{\text{ENTER}}$. The readout is .1221806653, so $\csc^{-1} 8.205 \approx 0.1222$ radians. We also use reciprocal relationships to find function values for arcsecant and arccotangent.

To find inverse function values in degrees, set the calculator in degree mode. Then use the keystrokes above to find that $\cos^{-1}(-0.2689) \approx 105.6°$ and $\csc^{-1} 8.205 \approx 7.0°$.

SINE REGRESSION

The SinReg operation on the TI-83 Plus and the TI-84 Plus can be used to fit a sine curve $y = a\sin(bx + c) + d$ to a set of data. At least four data points are required and there must be at least two data points per period. The output of SinReg is always in radians, regardless of the Radian/Degree mode setting. To see the graph, however, we must set the calculator in Radian mode.

The operations of entering data, making scatterplots, and graphing and evaluating the regression function are the same as for linear regression functions. Reread the material on pages 13 - 16 of this manual to review these procedures.

Section 7.5, Exercise 51 (a) Sales of certain products fluctuate in cycles. The data in the following table show the total sales of skis per month for a business in a northern climate.

Month, x	Total Sales, y (in thousands)
August, 8	$ 0
November, 11	7
February, 2	14
May, 5	7
August, 8	0

Using the sine regression feature on a graphing calculator, fit a sine function of the form $y = A\sin(Bx - C) + D$ to this set of data.

Enter the data in L_1 and L_2 as described in Chapter 1 of this manual. Press $\boxed{\text{STAT}}$ $\boxed{\triangleright}$ to view the STAT CALC menu. Then select SinReg by pressing $\boxed{\text{ALPHA}}$ C. If the data is entered in a combination of lists other than L_1 and L_2 their names must be entered separated by a comma. Since we have used L_1 and L_2 this is not necessary in this case. We also have the usual option of specifying a y = variable to which the regression equation can be stored. To select y_1, for example, press $\boxed{\text{VARS}}$ $\boxed{\triangleright}$ 1 1. Now press $\boxed{\text{ENTER}}$ to see the coefficients a, b, c, and d of the sine regression function $y = a\sin(bx + c) + d$

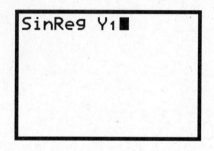

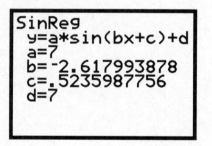

Chapter 8
Applications of Trigonometry

FINDING TRIGONOMETRIC NOTATION FOR COMPLEX NUMBERS

The TI-83 Plus and the TI-84 Plus can be used to find trigonometric notation for a complex number.

Section 8.3, Example 3 (a) Find trigonometric notation for $1 + i$.

Trigonometric notation for a complex number has the form $r(\cos\theta + i\sin\theta)$. We can find r using the abs feature from the MATH CPX menu. Press $\boxed{\text{MATH}}$ $\boxed{\triangleright}$ $\boxed{\triangleright}$ to display this menu. Then press 5 to copy "abs" to the home screen. (We could also use $\boxed{\triangledown}$ to highlight 5 and then press $\boxed{\text{ENTER}}$.) Then press 1 $\boxed{+}$ $\boxed{\text{2nd}}$ $\boxed{i}$ $\boxed{)}$ $\boxed{\text{ENTER}}$. The calculator returns $|1 + i|$, the value of r. It is approximately 1.414213562. This is a decimal approximation for $\sqrt{2}$.

Now use the MATH CPX menu again to find θ in degrees. First select Degree mode. Then press $\boxed{\text{MATH}}$ $\boxed{\triangleright}$ $\boxed{\triangleright}$ to display the MATH CPX menu. Select item 4, "angle," by pressing 4 or by using $\boxed{\triangledown}$ to highlight 4 and then pressing $\boxed{\text{ENTER}}$. Then press 1 $\boxed{+}$ $\boxed{\text{2nd}}$ $\boxed{i}$ $\boxed{)}$ $\boxed{\text{ENTER}}$. The calculator returns 45, so the angle θ is 45°. We can use the same procedure to find θ in radians after Radian mode has been selected.

```
abs(1+i)
        1.414213562
angle(1+i)
                45
■
```

CONVERTING FROM RECTANGULAR TO POLAR COORDINATES

The calculator can be used to convert from rectangular to polar coordinates, expressing the result using either degrees or radians. The calculator will supply a positive value for r and an angle in the interval $(-180°, 180°]$, or $(-\pi, \pi]$.

Section 8.4, Example 2 (a) Convert (3, 3) to polar coordinates.

To find r, regardless of the type of angle measure, press $\boxed{\text{2nd}}$ $\boxed{\text{ANGLE}}$ 5 3 $\boxed{,}$ 3 $\boxed{)}$ $\boxed{\text{ENTER}}$. The readout is 4.242640687, so $r \approx 4.2426$. This is a decimal approximation for $3\sqrt{2}$. Now, to find θ in degrees, set the calculator in Degree mode and press $\boxed{\text{2nd}}$ $\boxed{\text{ANGLE}}$ 6 3 $\boxed{,}$ 3 $\boxed{)}$ $\boxed{\text{ENTER}}$. The readout is 45, so $\theta = 45°$. Thus polar notation for (3,3) is (4.2426, 45°).

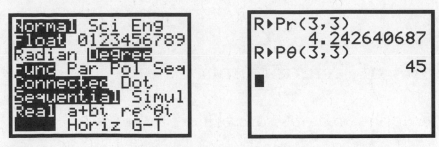

Set the calculator in Radian mode to find θ in radians. Repeat the keystrokes for finding θ above to find that $\theta \approx 0.7854$.

This is a decimal approximation for $\pi/4$. Thus polar notation for (3,3) can also be expressed as (4.2426,0.7854).

CONVERTING FROM POLAR TO RECTANGULAR COORDINATES

A graphing calculator can be used to convert from polar to rectangular coordinates.

Section 8.4, Example 3 Convert each of the following to rectangular coordinates.

(a) $(10, \pi/3)$ (b) $(-5, 135°)$

(a) Since the angle is given in radians, set the calculator in Radian mode. To find the x-coordinate of rectangular notation, press 2nd ANGLE 7 1 0 , 2nd π ÷ 3) ENTER. The readout is 5, so $x = 5$. The y-coordinate is found by pressing 2nd ANGLE 8 1 0 , 2nd π ÷ 3) ENTER. The readout is 8.660254038, so $y \approx 8.6603$. This is a decimal approximation of $5\sqrt{3}$. Thus, rectangular notation for $(10, \pi/3)$ is (5,8.6603).

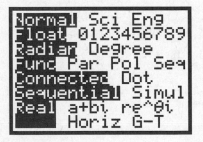

(b) The angle is given in degrees, so we set the calculator in Degree mode. To find the x-coordinate of rectangular notation, press 2nd ANGLE 7 (−) 5 , 1 3 5) ENTER. The readout is 3.535533906, so $x \approx 3.5355$. This is a decimal approximation of $\dfrac{5\sqrt{2}}{2}$. The y-coordinate is found by pressing 2nd ANGLE 8 (−) 5 , 1 3 5) ENTER. The readout is −3.535533906, so $y \approx -3.5355$. This is a decimal approximation of $-\dfrac{5\sqrt{2}}{2}$. Thus, rectangular notation for $(-5,135°)$ is (3.5355, −3.5355).

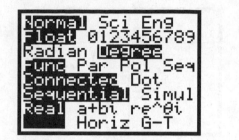

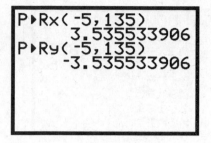

GRAPHING POLAR EQUATIONS

Polar equations can be graphed in either Radian mode or Degree mode. The equation must be written in the form $r = f(\theta)$ and the calculator must be set in Polar (Pol) mode. Typically we begin with a range of $[0, 2\pi]$ or $[0°, 360°]$, but it might be necessary to increase the range to ensure that sufficient points are plotted to display the entire graph.

Section 8.4, Example 6 Graph: $r = 1 - \sin\theta$.

First set the calculator in Polar mode by pressing $\boxed{\text{MODE}}$ $\boxed{\triangledown}$ $\boxed{\triangledown}$ $\boxed{\triangledown}$ $\boxed{\triangleright}$ $\boxed{\triangleright}$ $\boxed{\text{ENTER}}$. We will also select Radian mode.

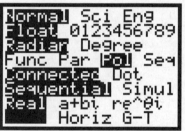

The equation is given in $r = f(\theta)$ form. Press $\boxed{\text{Y} =}$ to enter it on the "Y =" screen. Clear any existing entries and, with the cursor beside "$r_1 =$," press 1 $\boxed{-}$ $\boxed{\text{SIN}}$ $\boxed{\text{X,T,}\theta,n}$. Now press $\boxed{\text{WINDOW}}$ and enter the following settings:

$\theta\text{min} = 0$ (Smallest value of θ to be evaluated)
$\theta\text{max} = 2\pi$ (Largest value of θ to be evaluated)
$\theta\text{step} = \pi/24$ (Increment in θ values)
$\text{Xmin} = -3$
$\text{Xmax} = 3$
$\text{Xscl} = 1$
$\text{Ymin} = -3$
$\text{Ymax} = 1$
$\text{Yscl} = 1$

With these settings the calculator evaluates the function from $\theta = 0$ to $\theta = 2\pi$ in increments of $\pi/24$ and displays the graph in the square window $[-3, 3, -3, 1]$. Values entered in terms of π appear on the window screen as decimal approximations. Press $\boxed{\text{GRAPH}}$ to display the graph.

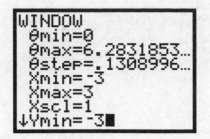 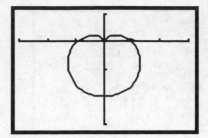

The curve can be traced with either rectangular or polar coordinates being displayed. The value of θ is also displayed when rectangular coordinates are selected. The choice of coordinates is made on the FORMAT screen. Press $\boxed{\text{2nd}}$ $\boxed{\text{FORMAT}}$ to display this screen. (FORMAT is the second operation associated with the $\boxed{\text{ZOOM}}$ key.) Then position the blinking cursor over RectGC to select rectangular coordinates or over PolarGC to select polar coordinates and press $\boxed{\text{ENTER}}$.

Chapter 9
Systems of Equations and Matrices

QUADRATIC REGRESSION

Quadratic functions can be fit to data using the quadratic regression operation from the STAT CALC menu. The operations of entering data, making scatterplots, and graphing and evaluating quadratic regression functions are the same as for linear regression functions.

Section 9.2, Exercise 37 *Morning Newspapers* The number of morning newspapers in the United States in various years is shown in the following table.

Year	Number of Morning Newspapers
1920	437
1940	380
1960	312
1980	387
2000	766
2004	813

(a) Use a graphing calculator to fit a quadratic function to the data, where x is the number of years after 1920.

(b) Use the function found in part (a) to estimate the number of morning newspapers in 2008.

(a) Clear any existing entries on the equation-editor screen. Then enter the data in L_1 and L_2 as described on page 14 of this manual. To fit a quadratic function to the data, press $\boxed{\text{STAT}}$ $\boxed{\triangleright}$ to view the STAT CALC menu. Then select QuadReg by pressing 5 $\boxed{\text{ENTER}}$. The coefficients of a quadratic equation $y = ax^2 + bx + c$ are displayed. Note that at least three data points are required for quadratic regression.

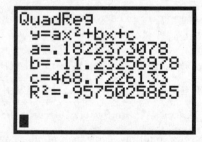

(b) To estimate the number of morning newspapers in 2008, we evaluate the regression function for $x = 88$. We can use any of the methods for evaluating a function found on pages 12 and 13 of this manual. On the next page we show a table set in Ask mode. We estimate that there were about 892 morning newspapers in 2008.

MATRICES AND ROW-EQUIVALENT OPERATIONS

Matrices with up to 99 rows or columns can be entered on a graphing calculator. As many as ten matrices can be entered at one time. Row-equivalent operations can be performed on matrices on the calculator.

Section 9.3, Example 1 Solve the following system:

$$2x - y + 4z = -3,$$
$$x - 2y - 10z = -6,$$
$$3x \quad + 4z = 7.$$

First we will enter the augmented matrix

$$\begin{bmatrix} 2 & -1 & 4 & -3 \\ 1 & -2 & -10 & -6 \\ 3 & 0 & 4 & 7 \end{bmatrix}$$

in the graphing calculator. Begin by pressing $\boxed{\text{2nd}}$ $\boxed{\text{MATRIX}}$ $\boxed{\triangleright}$ $\boxed{\triangleright}$ to display the MATRIX EDIT menu. (MATRIX is the second operation associated with the $\boxed{x^{-1}}$ key. On the TI-83 Plus this operation is spelled MATRX.) Then select the matrix to be defined. We will select matrix [**A**] by pressing 1. Now the MATRIX EDIT screen appears. The dimensions of the matrix are displayed on the top line of this screen, with the cursor on the row dimension. Enter the dimensions of the augmented matrix, 3 x 4, by pressing 3 $\boxed{\text{ENTER}}$ 4 $\boxed{\text{ENTER}}$. Now the cursor moves to the element in the first row and first column of the matrix. Enter the elements of the first row by pressing 2 $\boxed{\text{ENTER}}$ $\boxed{(-)}$ 1 $\boxed{\text{ENTER}}$ 4 $\boxed{\text{ENTER}}$ $\boxed{(-)}$ 3 $\boxed{\text{ENTER}}$. The cursor moves to the element in the second row and first column of the matrix. Enter the elements of the second and third rows of the augmented matrix by typing each in turn followed by $\boxed{\text{ENTER}}$ as above. Note that the screen displays only three columns of the matrix. The arrow keys can be used to move the cursor to any element at any time.

Row-equivalent operations are performed by making selections from the MATRIX MATH menu. To view this menu press $\boxed{\text{2nd}}$ $\boxed{\text{QUIT}}$ to leave the MATRIX EDIT screen. Then press $\boxed{\text{2nd}}$ $\boxed{\text{MATRIX}}$ $\boxed{\triangleright}$. Now press $\boxed{\triangledown}$ fifteen times or $\boxed{\triangle}$ one time to see the four row-equivalent operations, C: rowSwap(, D: row+(, E: *row(, and F: *row+(. These operations interchange two rows of a matrix, add two rows, multiply a row by a number, and multiply a row by a number and add it to a second row, respectively.

We will use the calculator to perform the row-equivalent operations that were done algebraically in the text. First, to interchange row 1 and row 2 of matrix [**A**], with the MATRIX MATH menu displayed, press $\boxed{\text{ALPHA}}$ $\boxed{\text{C}}$ to select rowSwap. Then press $\boxed{\text{2nd}}$ $\boxed{\text{MATRIX}}$ 1 to select [**A**]. Follow this with a comma and the rows to be interchanged, $\boxed{,}$ 1 $\boxed{,}$ 2 $\boxed{)}$ $\boxed{\text{ENTER}}$.

```
rowSwap([A],1,2)
 [[1  -2  -10  -6]
  [2  -1   4   -3]
  [3   0   4    7]]
■
```

The graphing calculator will not store the matrix produced using a row-equivalent operation, so when several operations are to be performed in succession it is helpful to store the result of each operation as it is produced. For example, to store the matrix resulting from interchanging the first and second rows of [**A**] as matrix [**B**], press $\boxed{\text{STO}\triangleright}$ $\boxed{\text{2nd}}$ $\boxed{\text{MATRIX}}$ 2 $\boxed{\text{ENTER}}$ immediately after interchanging the rows. This can also be done before $\boxed{\text{ENTER}}$ is pressed at the end of the rowSwap.

Next we multiply the first row of [**B**] by -2, add it to the second row and store the result as [**B**] again by pressing $\boxed{\text{2nd}}$ $\boxed{\text{MATRIX}}$ $\boxed{\triangleright}$ $\boxed{\text{ALPHA}}$ $\boxed{\text{F}}$ $\boxed{(-)}$ 2 $\boxed{,}$ $\boxed{\text{2nd}}$ $\boxed{\text{MATRIX}}$ 2 $\boxed{,}$ 1 $\boxed{,}$ 2 $\boxed{)}$ $\boxed{\text{STO}\triangleright}$ $\boxed{\text{2nd}}$ $\boxed{\text{MATRIX}}$ 2 $\boxed{\text{ENTER}}$. These keystrokes select *row+(from the MATRIX MATH menu; then they specify that the value of the multiplier is -2, the matrix being operated on is [**B**], and that a multiple of row 1 is being added to row 2; finally they store the result as [**B**].

To multiply row 1 by -3, add it to row 3, and store the result as [**B**], press $\boxed{\text{2nd}}$ $\boxed{\text{MATRIX}}$ $\boxed{\triangleright}$ $\boxed{\text{ALPHA}}$ $\boxed{\text{F}}$ $\boxed{(-)}$ 3 $\boxed{,}$ $\boxed{\text{2nd}}$ $\boxed{\text{MATRIX}}$ 2 $\boxed{,}$ 1 $\boxed{,}$ 3 $\boxed{)}$ $\boxed{\text{STO}\triangleright}$ $\boxed{\text{2nd}}$ $\boxed{\text{MATRIX}}$ 2 $\boxed{\text{ENTER}}$.

```
*row+(-3,[B],1,3
)→[B]
 [[1  -2  -10  -6]
  [0   3   24   9 ]
  [0   6   34   25]]
```

Now multiply the second row by $1/3$ and store the result as [**B**] again. Press $\boxed{\text{2nd}}$ $\boxed{\text{MATRIX}}$ $\boxed{\triangleright}$ $\boxed{\text{ALPHA}}$ $\boxed{\text{E}}$ 1 $\boxed{(\div)}$ 3 $\boxed{,}$ $\boxed{\text{2nd}}$ $\boxed{\text{MATRIX}}$ 2 $\boxed{,}$ 2 $\boxed{)}$ $\boxed{\text{STO}\triangleright}$ $\boxed{\text{2nd}}$ $\boxed{\text{MATRIX}}$ 2 $\boxed{\text{ENTER}}$. These keystrokes select *row(from the MATRIX MATH menu; then they specify that the value of the multiplier is $1/3$, the matrix being operated on is [**B**], and row 2 is being multiplied; finally they store the result as [**B**]. The keystrokes 1 $\boxed{(\div)}$ 3 could be replaced with 3 $\boxed{x^{-1}}$.

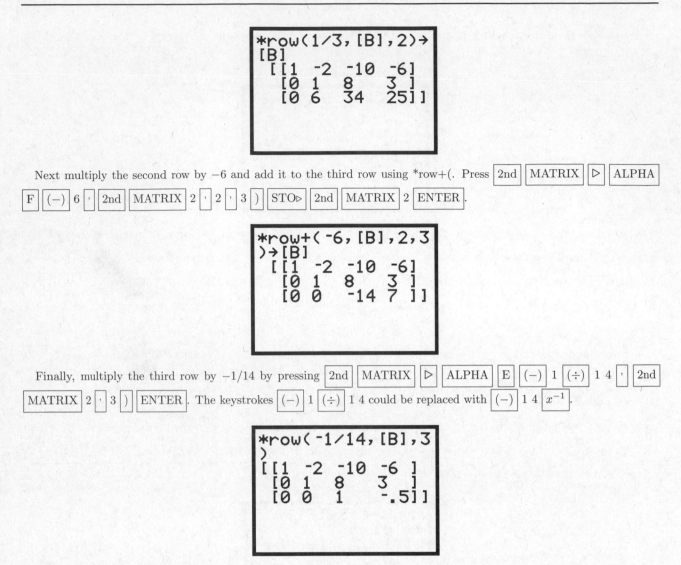

```
*row(1/3,[B],2)→
[B]
 [[1  -2  -10  -6]
  [0  1   8   3 ]
  [0  6   34  25]]
```

Next multiply the second row by -6 and add it to the third row using *row+(. Press 2nd MATRIX ▷ ALPHA F (−) 6 , 2nd MATRIX 2 , 2 , 3) STO▷ 2nd MATRIX 2 ENTER .

```
*row+( -6,[B],2,3
)→[B]
 [[1  -2  -10  -6]
  [0  1   8   3 ]
  [0  0  -14   7 ]]
```

Finally, multiply the third row by $-1/14$ by pressing 2nd MATRIX ▷ ALPHA E (−) 1 (÷) 1 4 , 2nd MATRIX 2 , 3) ENTER . The keystrokes (−) 1 (÷) 1 4 could be replaced with (−) 1 4 x^{-1}.

```
*row( -1/14,[B],3
)
 [[1  -2  -10  -6 ]
  [0  1   8   3  ]
  [0  0   1   -.5]]
```

Write the system of equations that corresponds to the final matrix. Then use back-substitution to solve for x, y, and z as illustrated in the text.

Instead of stopping with row-echelon form as we did above, we can continue to apply row-equivalent operations until the matrix is in reduced row-echelon form as in **Example 3** in **Section 9.3** of the text. Reduced row-echelon form of a matrix can be found directly by using the rref(operation from the MATRIX MATH menu. For example, to find reduced row-echelon form for matrix **A** in Example 1 above, after entering [**A**] and leaving the MATRIX EDIT screen, press 2nd MATRIX ▷ ALPHA B 2nd MATRIX 1 ENTER . We can read the solution of the system of equations, $(3, 7, -0.5)$ directly from the resulting matrix.

```
rref([A]
   [[1 0 0 3  ]
    [0 1 0 7  ]
    [0 0 1 -.5]]
```

USING THE PolySmlt APPLICATION

The application PolySmlt from the APPS menu can be used to solve a system of equations. If your calculator does not have this application pre-loaded, you might be able to download it from the Texas Instruments website, http://education.ti.com, or have it transmitted to your calculator from a calculator that contains the App.

We will do **Example 1** in **Section 9.3** again using the PolySmlt App.

Solve the following system:

$$2x - y + 4z = -3,$$
$$x - 2y - 10z = -6,$$
$$3x + 4z = 7.$$

To access PolySmlt, press the APPS key, scroll down to PolySmlt, and press ENTER. Next press any key to display the Main Menu. Press 2 to select item 2, SimultEqnSolve. Press 3 ENTER to enter the number of equations, and then press 3 ENTER again to enter the number of unknowns, or variables. The calculator displays a template for an augmented matrix. Enter the augmented matrix for this system of equations as described on page 50 of this manual. Finally, press the key directly below the "SOLVE" option at the bottom of the screen. This is the GRAPH key. The solution $(3, 7, -0.5)$ is displayed.

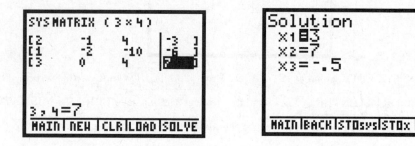

MATRIX OPERATIONS

We can use a graphing calculator to add and subtract matrices, to multiply a matrix by a scalar, and to multiply matrices.

Section 9.4, Example 1 (a) Find $\mathbf{A} + \mathbf{B}$ for

$$\mathbf{A} = \begin{bmatrix} -5 & 0 \\ 4 & \frac{1}{2} \end{bmatrix}, \ \mathbf{B} = \begin{bmatrix} 6 & -3 \\ 2 & 3 \end{bmatrix}.$$

Enter $\mathbf{A}$ and $\mathbf{B}$ on the MATRIX EDIT screen as $[\mathbf{A}]$ and $[\mathbf{B}]$ as described earlier in this chapter of the Graphing Calculator Manual. Press $\boxed{\text{2nd}}$ $\boxed{\text{QUIT}}$ to leave this screen. Then press $\boxed{\text{2nd}}$ $\boxed{\text{MATRIX}}$ $\boxed{1}$ $\boxed{+}$ $\boxed{\text{2nd}}$ $\boxed{\text{MATRIX}}$ $\boxed{2}$ $\boxed{\text{ENTER}}$ to display the sum.

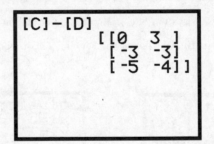

Section 9.4, Example 2 Find $\mathbf{C} - \mathbf{D}$ for each of the following.

a) $\mathbf{C} = \begin{bmatrix} 1 & 2 \\ -2 & 0 \\ -3 & -1 \end{bmatrix}, \ \mathbf{D} = \begin{bmatrix} 1 & -1 \\ 1 & 3 \\ 2 & 3 \end{bmatrix}$ 　　b) $\mathbf{C} = \begin{bmatrix} 5 & -6 \\ -3 & 4 \end{bmatrix}, \ \mathbf{D} = \begin{bmatrix} -4 \\ 1 \end{bmatrix}$

a) Enter $\mathbf{C}$ and $\mathbf{D}$ on the MATRIX EDIT screen as $[\mathbf{C}]$ and $[\mathbf{D}]$. Press $\boxed{\text{2nd}}$ $\boxed{\text{QUIT}}$ to leave this screen. Then press $\boxed{\text{2nd}}$ $\boxed{\text{MATRIX}}$ $\boxed{3}$ $\boxed{-}$ $\boxed{\text{2nd}}$ $\boxed{\text{MATRIX}}$ $\boxed{4}$ $\boxed{\text{ENTER}}$ to display the difference.

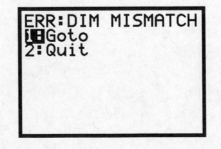

b) Enter $\mathbf{C}$ and $\mathbf{D}$ on the MATRIX EDIT screen as $[\mathbf{C}]$ and $[\mathbf{D}]$. Press $\boxed{\text{2nd}}$ $\boxed{\text{QUIT}}$ to leave this screen. Then press $\boxed{\text{2nd}}$ $\boxed{\text{MATRIX}}$ $\boxed{3}$ $\boxed{-}$ $\boxed{\text{2nd}}$ $\boxed{\text{MATRIX}}$ $\boxed{4}$ $\boxed{\text{ENTER}}$. The calculator returns the message ERR:DIM MISMATCH, indicating that this subtraction is not possible. This is the case because the matrices have different orders.

Section 9.4, Example 4 Find 3**A** and (-1)**A**, for $\mathbf{A} = \begin{bmatrix} -3 & 0 \\ 4 & 5 \end{bmatrix}$.

Enter **A** on the MATRIX EDIT screen as [A]. Press $\boxed{\text{2nd}}$ $\boxed{\text{QUIT}}$ to leave this screen. Then to find 3**A** press 3 $\boxed{\text{2nd}}$ $\boxed{\text{MATRIX}}$ 1 $\boxed{\text{ENTER}}$ and to find (-1)**A** press $\boxed{(-)}$ 1 $\boxed{\text{2nd}}$ $\boxed{\text{MATRIX}}$ 1 $\boxed{\text{ENTER}}$. Note that (-1)**A** is the opposite, or additive inverse, of **A** and can also be found by pressing $\boxed{(-)}$ $\boxed{\text{2nd}}$ $\boxed{\text{MATRIX}}$ 1 $\boxed{\text{ENTER}}$.

```
3[A]
         [[-9  0 ]
          [12 15]]
-1[A]
         [[3   0 ]
          [-4  -5]]
■
```

Section 9.4, Example 6 (a), (d) For

$$\mathbf{A} = \begin{bmatrix} 3 & 1 & -1 \\ 2 & 0 & 3 \end{bmatrix}, \mathbf{B} = \begin{bmatrix} 1 & 6 \\ 3 & -5 \\ -2 & 4 \end{bmatrix}, \text{ and } \mathbf{C} = \begin{bmatrix} 4 & -6 \\ 1 & 2 \end{bmatrix}$$

find each of the following.

a) AB   **d) AC**

First enter **A**, **B**,and **C** as [A], [B], and [C] on the MATRIX EDIT screen. Press $\boxed{\text{2nd}}$ $\boxed{\text{QUIT}}$ to leave this screen.

a) To find **AB** press $\boxed{\text{2nd}}$ $\boxed{\text{MATRIX}}$ 1 $\boxed{\text{2nd}}$ $\boxed{\text{MATRIX}}$ 2 $\boxed{\text{ENTER}}$.

```
[A][B]
        [[8   9 ]
         [-4 24]]
```

d) To find **AC** press $\boxed{\text{2nd}}$ $\boxed{\text{MATRIX}}$ 1 $\boxed{\text{2nd}}$ $\boxed{\text{MATRIX}}$ 3 $\boxed{\text{ENTER}}$. The calculator returns the message ERR:DIM MISMATCH, indicating that this multiplication is not possible. This is the case because the number of columns in **A** is not the same as the number of rows in **C**. Thus, the matrices cannot be multiplied in this order.

FINDING THE INVERSE OF A MATRIX

The inverse of a matrix can be found quickly on a graphing calculator.

Section 9.5, Example 3 Find $\mathbf{A}^{-1}$, where

$$\mathbf{A} = \begin{bmatrix} -2 & 3 \\ -3 & 4 \end{bmatrix}.$$

Enter $\mathbf{A}$ as [$\mathbf{A}$] on the MATRIX EDIT screen. Then press $\boxed{\text{2nd}}$ $\boxed{\text{QUIT}}$ to leave this screen. Now press $\boxed{\text{2nd}}$ $\boxed{\text{MATRIX}}$ 1 $\boxed{x^{-1}}$ $\boxed{\text{ENTER}}$ to display $\mathbf{A}^{-1}$.

```
[A]-1
         [[4  -3]
          [3  -2]]
```

Section 9.5, Exercise 7 Find $\mathbf{A}^{-1}$, where

$$\mathbf{A} = \begin{bmatrix} 6 & 9 \\ 4 & 6 \end{bmatrix}.$$

Enter $\mathbf{A}$ as [$\mathbf{A}$] on the MATRIX EDIT screen and then press $\boxed{\text{2nd}}$ $\boxed{\text{QUIT}}$ to leave this screen. Now press $\boxed{\text{2nd}}$ $\boxed{\text{MATRIX}}$ 1 $\boxed{x^{-1}}$ $\boxed{\text{ENTER}}$. The calculator returns the message ERR:SINGULAR MAT, indicating that $\mathbf{A}^{-1}$ does not exist.

MATRIX SOLUTIONS OF SYSTEMS OF EQUATIONS

We can write a system of n linear equations in n variables as a matrix equation $\mathbf{AX} = \mathbf{B}$. If $\mathbf{A}$ has an inverse the solution of the system of equations is given by $\mathbf{X} = \mathbf{A}^{-1}\mathbf{B}$.

Section 9.5, Example 5 Use an inverse matrix to solve the following system of equations:

$$-2x + 3y = 4,$$
$$-3x + 4y = 5.$$

Enter $\mathbf{A} = \begin{bmatrix} -2 & 3 \\ -3 & 4 \end{bmatrix}$ and $\mathbf{B} = \begin{bmatrix} 4 \\ 5 \end{bmatrix}$ on the MATRIX EDIT screen as [$\mathbf{A}$] and [$\mathbf{B}$]. Press $\boxed{\text{2nd}}$ $\boxed{\text{QUIT}}$ to leave this screen. Then press $\boxed{\text{2nd}}$ $\boxed{\text{MATRIX}}$ 1 $\boxed{x^{-1}}$ $\boxed{\text{2nd}}$ $\boxed{\text{MATRIX}}$ 2 $\boxed{\text{ENTER}}$. The result is the 2 x 1 matrix $\begin{bmatrix} 1 \\ 2 \end{bmatrix}$, so the solution is $(1, 2)$.

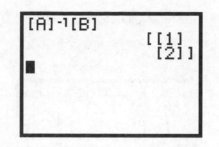

DETERMINANTS AND CRAMER'S RULE

We can evaluate determinants on a graphing calculator and use Cramer's rule to solve systems of equations.

Section 9.6, Example 5 Use a graphing calculator to evaluate $|\mathbf{A}|$.

$$\mathbf{A} = \begin{bmatrix} 1 & 6 & -1 \\ -3 & -5 & 3 \\ 0 & 4 & 2 \end{bmatrix}$$

Enter $\mathbf{A}$ on the MATRIX EDIT screen and then press $\boxed{\text{2nd}}$ $\boxed{\text{QUIT}}$ to leave this screen. We will select the "det(" operation from the MATRIX MATH menu. Press $\boxed{\text{2nd}}$ $\boxed{\text{MATRX}}$ $\boxed{\triangleright}$ 1 $\boxed{\text{2nd}}$ $\boxed{\text{MATRX}}$ 1 $\boxed{)}$ $\boxed{\text{ENTER}}$. We see that $|\mathbf{A}| = 26$.

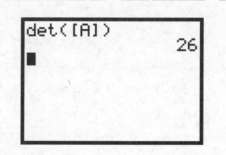

Section 9.6, Example 6 Solve using Cramer's rule:

$$2x + 5y = 7,$$
$$5x - 2y = -3.$$

First we enter the matrices corresponding to D, D_x, and D_y as $\mathbf{A}$, $\mathbf{B}$, and $\mathbf{C}$, respectively. We have

$$\mathbf{A} = \begin{bmatrix} 2 & 5 \\ 5 & -2 \end{bmatrix}, \mathbf{B} = \begin{bmatrix} 7 & 5 \\ -3 & -2 \end{bmatrix}, \text{ and } \mathbf{C} = \begin{bmatrix} 2 & 7 \\ 5 & -3 \end{bmatrix}.$$

Then we use the "det(" operation from the MATRIX MATH menu. We have

$$x = \frac{\det(\mathbf{B})}{\det(\mathbf{A})} \text{ and } y = \frac{\det(\mathbf{C})}{\det(\mathbf{A})}.$$

To find x expressed in fraction notation, press $\boxed{\text{2nd}}$ $\boxed{\text{MATRIX}}$ $\boxed{\triangleright}$ 1 $\boxed{\text{2nd}}$ $\boxed{\text{MATRIX}}$ 2 $\boxed{)}$ $\boxed{\div}$ $\boxed{\text{2nd}}$ $\boxed{\text{MATRIX}}$ $\boxed{\triangleright}$ 1 $\boxed{\text{2nd}}$ $\boxed{\text{MATRIX}}$ 1 $\boxed{)}$ $\boxed{\text{MATH}}$ 1 $\boxed{\text{ENTER}}$. The result is $-\frac{1}{29}$. To find y expressed in fraction notation, press $\boxed{\text{2nd}}$ $\boxed{\text{MATRIX}}$ $\boxed{\triangleright}$ 1 $\boxed{\text{2nd}}$ $\boxed{\text{MATRIX}}$ 3 $\boxed{)}$ $\boxed{\div}$ $\boxed{\text{2nd}}$ $\boxed{\text{MATRIX}}$ $\boxed{\triangleright}$ 1 $\boxed{\text{2nd}}$ $\boxed{\text{MATRIX}}$ 1 $\boxed{)}$ $\boxed{\text{MATH}}$ 1 $\boxed{\text{ENTER}}$. We can also recall the entry for x and edit it, replacing matrix **B** with matrix **C** to find y. In either case the result is $\dfrac{41}{29}$. The solution of the system of equations is $\left(-\dfrac{1}{29}, \dfrac{41}{29}\right)$.

```
det([B])/det([A]
)▶Frac
               -1/29
det([C])/det([A]
)▶Frac
               41/29
■
```

GRAPHS OF INEQUALITIES

We can graph linear inequalities on a graphing calculator, shading the region of the solution set. The calculator should be set in Func mode at this point.

Section 9.7, Example 1 Graph: $y < x + 3$.

First we graph the related equation $y = x + 3$. We use the standard window $[-10, 10, -10, 10]$. Since the inequality symbol is $<$ we know that the line $y = x + 3$ is not part of the solution set. In a hand-drawn graph we would use a dashed line to indicate this. We might try selecting Dot mode or use the Dot GraphStyle, but even if Dot mode is used, the line appears to be solid. After determining that the solution set of the inequality consists of all points below the line, we use the calculator's SHADE operation to shade this region. SHADE is item 7 on the DRAW DRAW menu. To access it press $\boxed{\text{2nd}}$ $\boxed{\text{DRAW}}$ 7.

Now enter a lower function and an upper function. The region between them will be shaded. We want to shade the area between the bottom of the window, $y = -10$, and the line $y = x + 3$ so we enter $\boxed{(-)}$ $\boxed{1\,0}$ $\boxed{,}$ $\boxed{\text{X,T,}\Theta, n}$ $\boxed{+}$ $\boxed{3}$ $\boxed{)}$ $\boxed{\text{ENTER}}$. We can also enter $x + 3$ as y_1. The result is shown below. Keep in mind that the line $y = x + 3$ is not included in the solution set.

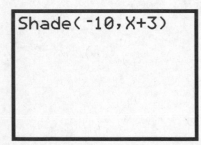
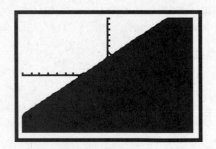

The "shade below" GraphStyle can also be used to shade this region. After entering the related equation, cycle through the Graph Style options on the "Y =" screen as described on page 20 of this manual until the "shade below" option appears. Then press $\boxed{\text{GRAPH}}$ to display the graph of the inequality. As mentioned above, keep in mind the fact that the line $y = x + 3$ is not included in the solution set.

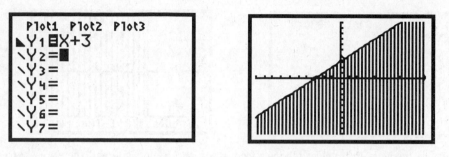

Some calculators have a pre-loaded application, Inequalz, on the APPS menu that can be used to graph inequalities. If your calculator does not have this App, you might be able to download it from the Texas Instruments website, http://education.ti.com, or have it transmitted to your calculator from a calculator that contains it. To use Inequalz to graph $y < x + 3$, first press the APPS key, scroll down to Inequalz, and press ENTER. Then press any key to display the Y = screen. Enter $y_1 < x + 3$ by first pressing ALPHA F2 to select the inequality symbol < from the menu at the bottom of the screen. (F2 is the Alpha operation associated with the WINDOW key. This key is used because it lies directly below the < symbol on the screen.)

Now press ▷ to position the cursor to the right of $Y_1 <$ and continue by pressing X, T, Θ, n + 3. Then select a window and press GRAPH to see the graph of the inequality. Here we pressed ZOOM 6 to select the standard window. Note that the App has the capability to graph the related equation, $y = x + 3$, as a dashed line.

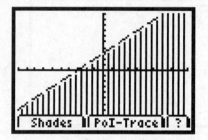

To quit the Inequalz App, select Inequalz from the APPS menu again and then press 2 to select Quit from the INEQUALZ RUNNING menu.

We can also use the Inequalz App to graph an inequality with a related equation of the form $x = a$.

Section 9.7, Example 3 Graph $x > -3$ on a plane.

Select the Inequalz App as described above. Then press △ to position the cursor on X = at the top of the screen and press ENTER. Then enter $X_1 > -3$ by first pressing ALPHA F4 to select the inequality symbol >. (F4 is the Alpha operation associated with the TRACE key. It is the key directly below the > symbol on the screen.) Next press ▷ to position the cursor to the right of $X_1 >$ and continue to enter the inequality by pressing (−) 3. Then select a window and press GRAPH to see the graph of the inequality. The standard window is shown below.

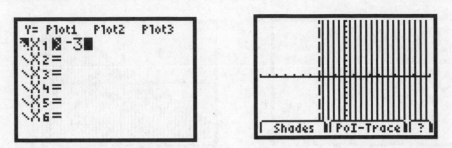

Recall that, to quit the Inequalz App, select Inequalz from the APPS menu again and then press 2 to select Quit.

We can use the SHADE operation to graph a system of inequalities when the solution set lies between the graphs of two functions.

Section 9.7, Exercise 43 Graph:
$$y \leq x,$$
$$y \geq 3 - x.$$

First graph the related equations $y_1 = x$ and $y_2 = 3 - x$ and determine that the solution set consists of all the points on or above the graph of $y_2 = 3 - x$ and on or below the graph of $y_1 = x$. We will shade this region by pressing $\boxed{\text{2nd}}$ $\boxed{\text{DRAW}}$ 7 3 $\boxed{-}$ $\boxed{\text{X,T,}\Theta\text{,}n}$ $\boxed{\text{,}}$ $\boxed{\text{X,T,}\Theta\text{,}n}$ $\boxed{\text{)}}$ $\boxed{\text{ENTER}}$. These keystrokes select the SHADE operation from the DRAW DRAW menu and then enter $y_2 = 3 - x$ as the lower function and $y_1 = x$ as the upper function. We could also enter these functions as y_2 and y_1, respectively.

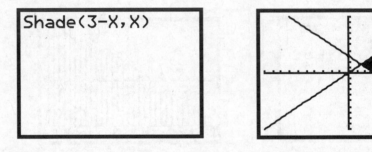

We can also graph systems of inequalities by shading the solution set of each inequality in the system with a different pattern. When the "shade above" or "shade below" GraphStyle options are selected the calculator rotates through four shading patterns. Vertical lines shade the first function, horizontal lines the second, negatively sloping diagonal lines the third, and positively sloping diagonal lines the fourth. These patterns repeat if more than four functions are graphed.

Section 9.7, Example 5 Graph the solution set of the system
$$x + y \leq 4,$$
$$x - y \geq 2.$$

First graph the equation $x + y = 4$, entering it in the form $y_1 = -x + 4$. We determine that the solution set of $x + y \leq 4$ consists of all points below the line $x + y = 4$, or $y_1 = -x + 4$, so we select the "shade below" GraphStyle for this function. Next graph $x - y = 2$, entering it in the form $y_2 = x - 2$. The solution set of $x - y \geq 2$ is all points below the line $x - y = 2$, or $y_2 = x - 2$, so we also choose the "shade below" GraphStyle for this function. Now press $\boxed{\text{GRAPH}}$ to display the solution sets of each inequality in the system and the region where they overlap. The region of overlap is the solution set of the system of inequalities.

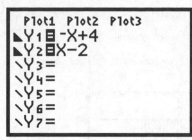
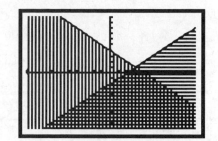

We can also use the Inequalz App to graph this system of inequalities. First we solve each inequality for y, obtaining $y \leq 4 - x$ from $x + y \leq 4$ and $y \leq x - 2$ from $x - y \geq 2$. Press $\boxed{\text{APPS}}$, select the Inequalz App, and enter $y_1 \leq 4 - x$ and $y_2 \leq x - 2$ as described on page 59 of this manual. Then select a viewing window and press $\boxed{\text{GRAPH}}$. Here we pressed $\boxed{\text{ZOOM}}$ 6 to select the standard window. Initially we see a graph like the one above, with the solution sets of both inequalities shaded. To show only the solution set of the system of inequalities, press $\boxed{\text{ALPHA}}$ $\boxed{\text{F1}}$ to select Shades. (F1 is the Alpha operation associated with the $\boxed{\text{Y} =}$ key.) Then press 1 to select Intersection and display the graph shown below.

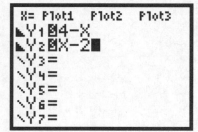
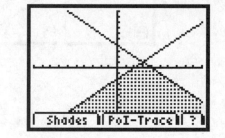

FINDING THE COORDINATES OF VERTICES

If a system of inequalities is graphed using a shading option, the coordinates of the vertices can be found using the Intersect feature. If a system of inequalities is graphed using the Inequalz application from the APPS menu, the PoI-Trace feature can be used to find the coordinates of the vertices.

Section 9.7, Example 6 Graph the following system of inequalities and find the coordinates of any vertices formed:

$$3x - y \leq 6, \qquad (1)$$
$$y - 3 \leq 0, \qquad (2)$$
$$x + y \geq 0. \qquad (3)$$

We will demonstrate the use of the PoI-Trace feature. First we access the Inequalz App and graph the three inequalities as described on page 59 of this manual. To do this we first solve each inequality for y and graph $y_1 \geq 3x - 6$, $y_2 \leq 3$, and $y_3 \geq -x$. In the graphs shown on the next page we have used the window $[-5, 5, -5, 5]$ and we have also used the Shades feature to shade only the solution set of the system of inequalities. The use of Shades is described in Example 5 above.

To find the coordinates of the vertices, first press $\boxed{\text{ALPHA}}$ $\boxed{\text{F3}}$ to select the PoI-Trace option at the bottom of the screen. (F3 is the Alpha option associated with the $\boxed{\text{ZOOM}}$ key.) We see the coordinates of the vertex (3,3) displayed at the bottom of the screen. The notation Y1∩Y2 at the top of the screen tells us that this is the vertex associated with the system of equations from inequalities (1) and (2) above. Press the $\boxed{\triangledown}$ key to display the coordinates of the vertex associated with inequalities (1) and (3), (1.5, −1.5), and press $\boxed{\triangleleft}$ to find the vertex associated with (2) and (3), (−3, 3).

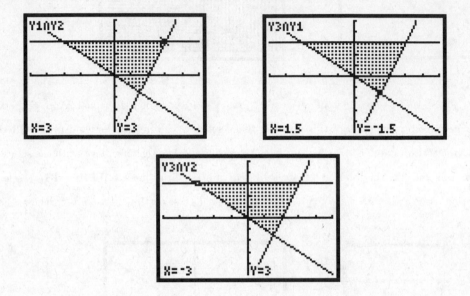

Chapter 10
Conic Sections

Many conic sections are represented by equations that are not functions. Consequently, these equations must be entered on the equation-editor screen as two equations, each of which is a function. Alternatively, some calculators have a pre-loaded application that allows conic sections to be graphed in standard form.

GRAPHING PARABOLAS

To graph a parabola of the form $y^2 = 4px$ or $(y - k)^2 = 4p(x - h)$, we must first solve the equation for y.

Section 10.1, Example 4 Graph the parabola $y^2 - 2y - 8x - 31 = 0$.

In the text we used the quadratic formula to solve the equation for y:
$$y = \frac{2 \pm \sqrt{32x + 128}}{2}.$$

One way to produce the graph of the parabola is to enter $y_1 = \dfrac{2 + \sqrt{32x + 128}}{2}$ and $y_2 = \dfrac{2 - \sqrt{32x + 128}}{2}$, select a window, and press $\boxed{\text{GRAPH}}$ to see the graph. Here we use the square window $[-12, 12, -8, 8]$. The first equation produces the top half of the parabola and the second equation produces the lower half.

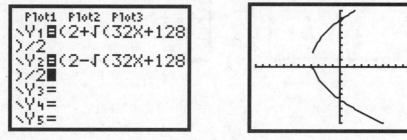

We can also enter $y_1 = \sqrt{32x + 128}$ and then enter $y_2 = \dfrac{2 + y_1}{2}$ and $y_3 = \dfrac{2 - y_1}{2}$. For example, to enter $y_2 = \dfrac{2 + y_1}{2}$ position the cursor beside "Y2 =" and press $\boxed{(}$ $\boxed{2}$ $\boxed{+}$ $\boxed{\text{VARS}}$ $\boxed{\triangleright}$ 1 1 $\boxed{)}$ $\boxed{\div}$ 2. Enter $y_3 = \dfrac{2 - y_1}{2}$ in a similar manner. Finally, deselect y_1 by moving the cursor to the equals sign following Y1 and pressing $\boxed{\text{ENTER}}$. Note that the equals sign in this equation is no longer highlighted. This indicates that it has been deselected and thus its graph will not appear with the graphs of the equations that remain selected.

The top half of the graph is produced by y_2 and the lower half by y_3. The expression for y_1 was entered to avoid entering the radical expression more than once. By deselecting y_1 we prevent its graph from appearing on the screen with the graph of the parabola.

We could also use the following form of the equation of the parabola found in the text:

$$(y - 1)^2 = 8(x + 4).$$

Solve this equation for y.

$$y - 1 = \pm\sqrt{8(x + 4)}$$
$$y = 1 \pm \sqrt{8(x + 4)}$$

Then enter $y_1 = 1 + \sqrt{8(x + 4)}$ and $y_2 = 1 - \sqrt{8(x + 4)}$, or enter $y_1 = \sqrt{8(x + 4)}$, $y_2 = 1 + y_1$, and $y_3 = 1 - y_1$, and deselect y_1 as described above.

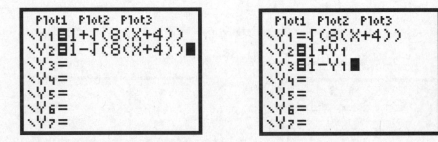

The Conics application from the APPS menu can also be used to graph a parabola. This App requires the equation to be written in standard form, $(y - k)^2 = 4p(x - h)$, or $(x - h)^2 = 4p(y - k)$. In the text we found standard form for the equation of the parabola $y^2 - 2y - 8x - 31 = 0$ to be $(y - 1)^2 = 4(2)[x - (-4)]$.

To use the Conics App to graph this parabola, first press $\boxed{\text{APPS}}$, scroll down to Conics, and then press $\boxed{\text{ENTER}}$ to see the Conics menu. Next press 4 to select Parabola and then press 1 or $\boxed{\text{ENTER}}$ to select an equation in the form $(Y - K)^2 = 4P(X - H)$. We have H = −4, K = 1, and P = 2. Enter these values on the parabola screen by pressing $\boxed{(-)}$ 4 $\boxed{\text{ENTER}}$ 1 $\boxed{\text{ENTER}}$ 2 $\boxed{\text{ENTER}}$. (Note that the vertical line to the right of the 2 on the screen on the left below is the cursor, not the number 1.) Finally press $\boxed{\text{GRAPH}}$ to graph the parabola in a window selected by the calculator. If you wish to select a window manually, press $\boxed{\text{MODE}}$ and select MAN. Then press $\boxed{\text{Y} =}$, which corresponds to the ESC (Escape) command, to leave the Mode screen. Finally, press $\boxed{\text{WINDOW}}$ and enter the desired dimensions.

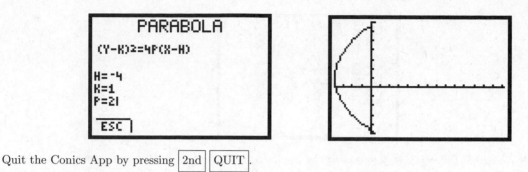

Quit the Conics App by pressing $\boxed{\text{2nd}}$ $\boxed{\text{QUIT}}$.

GRAPHING CIRCLES

The equation of a circle must be solved for y before it can be entered on the equation-editor screen. The Conics App can also be used to graph a circle, if available. We can also use the Circle feature from the DRAW menu, as discussed on page 11 of this manual.

Section 10.2, Example 1 Graph the circle $x^2 + y^2 - 16x + 14y + 32 = 0$.

In the text we found the standard form for the equation of the circle and then solved for y:

$$y = -7 \pm \sqrt{81 - (x-8)^2}.$$

We could also have solved the original equation using the quadratic formula.

One way to produce the graph is to enter $y_1 = -7 + \sqrt{81 - (x-8)^2}$ and $y_2 = -7 - \sqrt{81 - (x-8)^2}$, select a square window, and press $\boxed{\text{GRAPH}}$. Here we use $[-12, 24, -20, 4]$. The first equation produces the top half of the circle and the second equation produces the lower half.

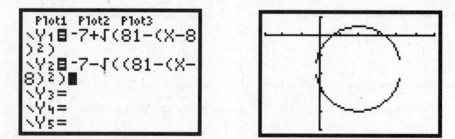

We can also enter $y_1 = \sqrt{81 - (x-8)^2}$ and then enter $y_2 = -7 + y_1$ and $y_3 = -7 - y_1$. Then deselect y_1, select a square window, and press $\boxed{\text{GRAPH}}$. We use y_1 to eliminate the need to enter the radical expression more than once. Deselecting it prevents the graph of y_1 from appearing on the screen with the graph of the circle. The top half of the graph is produced by y_2 and the lower half by y_3.

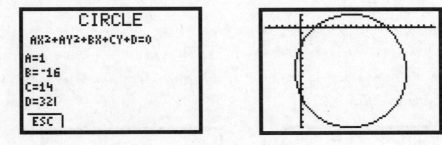

To graph the circle $x^2 + y^2 - 16x + 14y + 32 = 0$ using the Conics App, first press $\boxed{\text{APPS}}$, scroll down to Conics, and press $\boxed{\text{ENTER}}$ or 1 to select Circle. We see that we can enter an equation in standard form, $(X - H)^2 + (Y - K)^2 = R^2$ or in the form $AX^2 + AY^2 + BX + CY + D = 0$. Since our equation is given in the second form, we press 2 to select the second option.

We have $A = 1$, $B = -16$, $C = 14$, and $D = 32$. Enter these values by pressing 1 $\boxed{\text{ENTER}}$ $\boxed{(-)}$ 1 6 $\boxed{\text{ENTER}}$ 1 4 $\boxed{\text{ENTER}}$ 3 2 $\boxed{\text{ENTER}}$. (Note that the vertical line to the right of 32 on the screen on the left below is the cursor, not the number 1.) Finally, press $\boxed{\text{GRAPH}}$ to see the circle in a window selected by the calculator. The window can be selected manually as described in Example 1 above.

GRAPHING ELLIPSES

The equation of an ellipse must be solved for y before it can be entered on the equation-editor screen. The Conics App can also be used to graph an ellipse, if available. The procedure for graphing an ellipse of the form $\dfrac{x^2}{a^2} + \dfrac{y^2}{b^2} = 1$ or $\dfrac{x^2}{b^2} + \dfrac{y^2}{a^2} = 1$ is described in **Section 10.2, Example 2** in the text. The procedure for using the Conics App that follows on page 68 of this manual can also be used to graph ellipses in this form.

Now we consider ellipses of the form $\dfrac{(x-h)^2}{a^2} + \dfrac{(y-k)^2}{b^2} = 1$ or $\dfrac{(x-h)^2}{b^2} + \dfrac{(y-k)^2}{a^2} = 1$.

Section 10.2, Example 4 Graph the ellipse $4x^2 + y^2 + 24x - 2y + 21 = 0$.

Completing the square in the text, we found that the equation can be written as
$$\frac{(x+3)^2}{4} + \frac{(y-1)^2}{16} = 1.$$
Solve this equation for y.

$$\frac{(x+3)^2}{4} + \frac{(y-1)^2}{16} = 1$$

$$\frac{(y-1)^2}{16} = 1 - \frac{(x+3)^2}{4}$$

$$(y-1)^2 = 16 - 4(x+3)^2 \qquad \text{Multiplying by 16}$$

$$y - 1 = \pm\sqrt{16 - 4(x+3)^2}$$

$$y = 1 \pm \sqrt{16 - 4(x+3)^2}$$

Now we can use this equation to produce the graph in either of two ways. One is to enter $y_1 = 1 + \sqrt{16 - 4(x+3)^2}$ and $y_2 = 1 - \sqrt{16 - 4(x+3)^2}$, select a square window, and press $\boxed{\text{GRAPH}}$. Here we use $[-9, 9, -6, 6]$. The first equation produces the top half of the ellipse and the second equation produces the lower half.

We can also enter $y_1 = \sqrt{16 - 4(x+3)^2}$ and then enter $y_2 = 1 + y_1$ and $y_3 = 1 - y_1$. Deselect y_1, select a square window, and press $\boxed{\text{GRAPH}}$. We use y_1 to eliminate the need to enter the radical expression more than once. Deselecting it prevents the graph of y_1 from appearing on the screen with the graph of the ellipse. The top half of the graph is produced by y_2 and the lower half by y_3.

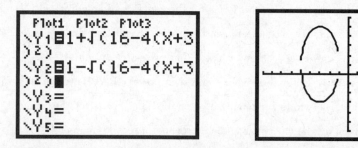

We could also begin by using the quadratic formula to solve the original equation for y.

$$4x^2 + y^2 + 24x - 2y + 21 = 0$$

$$y^2 - 2y + (4x^2 + 24x + 21) = 0$$

$$y = \frac{-(-2) \pm \sqrt{(-2)^2 - 4 \cdot 1 \cdot (4x^2 + 24x + 21)}}{2 \cdot 1}$$

$$y = \frac{2 \pm \sqrt{4 - 16x^2 - 96x - 84}}{2}$$

$$y = \frac{2 \pm \sqrt{-16x^2 - 96x - 80}}{2}$$

Then enter $y_1 = \frac{2 + \sqrt{-16x^2 - 96x - 80}}{2}$ and $y_2 = \frac{2 - \sqrt{-16x^2 - 96x - 80}}{2}$, or enter $y_1 = \sqrt{-16x^2 - 96x - 80}$, $y_2 = \frac{2 + y_1}{2}$, and $y_3 = \frac{2 - y_1}{2}$, and deselect y_1.

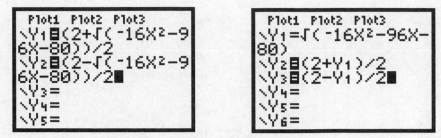

Select a square window and press $\boxed{\text{GRAPH}}$ to display the graph shown above.

To use the Conics App to graph the ellipse, first press $\boxed{\text{APPS}}$, scroll down to Conics, press $\boxed{\text{ENTER}}$, and then press 2 to select the Ellipse menu. That menu displays the standard forms for the equation of an ellipse with center at the origin, the first form for a horizontal major axis and the second for a vertical axis. Recall that our equation can be written as $\dfrac{[x-(-3)]^2}{2^2} + \dfrac{(y-1)^2}{4^2} = 1$, so this ellipse has a vertical major axis. Thus, we select option 2, $\dfrac{(X-H)^2}{B^2} + \dfrac{(Y-K)^2}{A^2} = 1$. We have A = 4, B = 2, H = −3, and K = 1. Enter these values and then press $\boxed{\text{GRAPH}}$ to see the graph in a window selected by the calculator. We can select the window manually as described on page 64 of this manual. (Note that the vertical line to the right of 1 on the screen on the left below is the cursor, not a second number 1.)

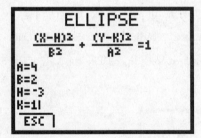
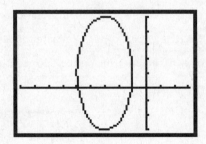

GRAPHING HYPERBOLAS

As with equations of circles, parabolas, and ellipses, equations of hyperbolas must be solved for y before they can be entered on the equation-editor screen. They can also be graphed using the Conics App, if it is available.

Section 10.3, Example 2 Graph the hyperbola $9x^2 - 16y^2 = 144$.

First solve the equation for y.

$$9x^2 - 16y^2 = 144$$
$$-16y^2 = -9x^2 + 144$$
$$y^2 = \frac{-9x^2 + 144}{-16}$$
$$y = \pm\sqrt{\frac{-9x^2 + 144}{-16}}, \text{ or } \pm\sqrt{\frac{9x^2 - 144}{16}}$$

It is not necessary to simplify further.

Now enter $y_1 = \sqrt{\dfrac{9x^2 - 144}{16}}$ and either $y_2 = -\sqrt{\dfrac{9x^2 - 144}{16}}$ or $y_2 = -y_1$, select a square window, and press $\boxed{\text{GRAPH}}$.

Here we use the window $[-9, 9, -6, 6]$. The top half of the graph is produced by y_1 and the lower half by y_2.

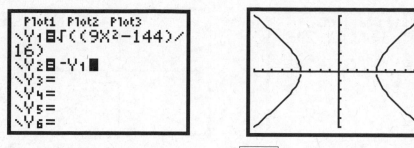

To use the Conics App to graph this hyperbola, first press APPS , scroll down to Conics, press ENTER , and then press 3 to display the Hyperbola menu. In the text we see that our equation can be written in the form $\dfrac{x^2}{4^2} - \dfrac{y^2}{3^2} = 1$, so we select option 1. We have A = 4, B = 3, H = 0, and K = 0. Enter these values and then press GRAPH to see the graph in a window selected by the calculator. The window can be selected manually as described on page 64 of this manual. (Note that the vertical line beside the last 0 on the screen on the left below is the cursor, not the number 1.)

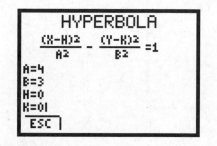 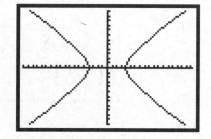

Section 10.3, Example 3 Graph the hyperbola $4y^2 - x^2 + 24y + 4x + 28 = 0$.

In the text we completed the square to get the standard form of the equation. Now solve the equation for y.

$$\frac{(y+3)^2}{1} - \frac{(x-2)^2}{4} = 1$$

$$(y+3)^2 = \frac{(x-2)^2}{4} + 1$$

$$y + 3 = \pm\sqrt{\frac{(x-2)^2}{4} + 1}$$

$$y = -3 \pm \sqrt{\frac{(x-2)^2}{4} + 1}$$

This equation can be used to produce the graph in either of two ways. One is to enter $y_1 = -3 + \sqrt{\dfrac{(x-2)^2}{4} + 1}$ and $y_2 = -3 - \sqrt{\dfrac{(x-2)^2}{4} + 1}$, select a square window, and press GRAPH . Here we use $[-12, 12, -9, 9]$. The first equation produces the top half of the hyperbola and the second the lower half.

 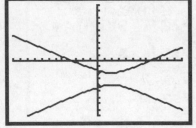

We can also enter $y_1 = \sqrt{\dfrac{(x-2)^2}{4} + 1}$, $y_2 = -3 + y_1$, and $y_3 = -3 - y_1$. Then deselect y_1, select a square window, and press $\boxed{\text{GRAPH}}$. Again, y_1 is used to eliminate the need to enter the radical expression more than once. Deselecting it prevents the graph of y_1 from appearing on the screen with the graph of the hyperbola. The top half of the graph is produced by y_2 and the lower half by y_3.

```
Plot1  Plot2  Plot3
\Y1=√((X-2)²/4+1
)
\Y2◻-3+Y1
\Y3◻-3-Y1■
\Y4=
\Y5=
\Y6=
```

If the Conics App is available, it can also be used to graph this hyperbola. Access the Hyperbola menu in the Conics App as described earlier in this section on graphing hyperbolas. We have $\dfrac{(y+3)^2}{1} - \dfrac{(x-2)^2}{4} = 1$, or $\dfrac{[y-(-3)]^2}{1^2} - \dfrac{(x-2)^2}{2^2} = 1$, so we select option 2 and enter A = 1, B = 2, H = 2, and K = −3. (Note that the vertical line beside −3 on the screen on the left below is the cursor, not the number 1.) Finally press $\boxed{\text{GRAPH}}$ to see the graph in a window selected by the calculator. We can select a window manually as described on page 64 of this manual. Here we have selected the window $[-12, 12, -9, 9]$.

```
     HYPERBOLA
   (Y-K)²   (X-H)²
   ------ - ------ =1
    A²       B²
A=1
B=2
H=2
K=-3|
 ESC
```

GRAPHING PARAMETRIC EQUATIONS

Plane curves described with parametric equations can be graphed on a graphing calculator.

Section 10.7, Example 2 (a) Using a graphing calculator, graph the plane curve given by the set of parametric equations and the restriction on the parameter.

$$x = t^2, \ y = t - 1,; \ -1 \leq t \leq 4$$

First press $\boxed{\text{MODE}}$ and select Parametric (Par) mode.

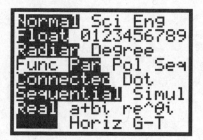

Then press $\boxed{\text{Y} =}$ to display the equation-editor screen. Enter $X_{1T} = t^2$ and $Y_{1T} = t - 1$. Note that, when the $\boxed{\text{X, T}, \theta, n}$ key is pressed in Parametric mode, the variable T is produced. Now press $\boxed{\text{WINDOW}}$ and enter the following settings:

Tmin = −1	(Smallest value of T to be evaluated)
Tmax = 4	(Largest value of T to be evaluated)
Tstep = .1	(Increment in T values)
Xmin = −2	
Xmax =18	
Xscl = 1	
Ymin = −4	
Ymax = 4	
Yscl = 1	

Since $x = t^2$ and $-1 \leq t \leq 4$, we have $0 \leq x \leq 16$. Thus, we choose Xmin and Xmax to display this interval. Similarly, since $y = t - 1$, we have $-2 \leq y \leq 3$ and we choose Ymin and Ymax to show this interval. Press $\boxed{\text{GRAPH}}$ to display the graph.

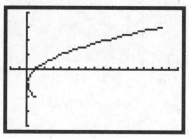

Chapter 11
Sequences, Series, and Combinatorics

The computational capabilities of a graphing calculator can be used when working with sequences, series, and combinatorics.

FINDING THE TERMS OF A SEQUENCE

Section 11.1, Example 2 Use a graphing calculator to find the first 5 terms of the sequence whose general term is given by $a_n = n/(n+1)$.

Although we could use a table, we will use the Seq feature. The calculator can be set in either Func or Seq mode when this feature is used. Press $\boxed{\text{2nd}}$ $\boxed{\text{LIST}}$ $\boxed{\triangleright}$ to display the LIST OPS menu. (LIST is the second operation associated with the $\boxed{\text{STAT}}$ key.) Then press 5 to paste "seq(" to the home screen. Then enter the general term of the sequence, replacing n with x if the calculator is in Func mode. Follow this with the variable and the numbers of the first and last terms desired. We will also use the ▷Frac feature to express the terms as fractions. Press $\boxed{\text{X, T, }\Theta, n}$ $\boxed{\div}$ $\boxed{(}$ $\boxed{\text{X, T, }\Theta, n}$ $\boxed{+}$ $\boxed{1}$ $\boxed{)}$ $\boxed{,}$ $\boxed{\text{X, T, }\Theta, n}$ $\boxed{,}$ 1 $\boxed{,}$ 5 $\boxed{)}$ $\boxed{\text{MATH}}$ $\boxed{\text{ENTER}}$ $\boxed{\text{ENTER}}$.

If the calculator is in Func mode, pressing $\boxed{\text{X, T, }\Theta, n}$ will produce the variable x in the expression. If it is in Sequence mode the variable will be n. Use the $\boxed{\triangleright}$ and $\boxed{\triangleleft}$ keys to move back and forth between the screens shown below.

FINDING PARTIAL SUMS

We can use a graphing calculator to find partial sums of a sequence when a formula for the general term is known.

Section 11.1, Example 6 Use a graphing calculator to find $S_1, S_2, S_3,$ and S_4 for the sequence whose general term is given by $a_n = n^2 - 3$.

We will use the cumSum feature. The calculator will write the partial sums as a list. First press $\boxed{\text{2nd}}$ $\boxed{\text{LIST}}$ $\boxed{\triangleright}$ 6 to paste "cumSum(" to the home screen. Then press $\boxed{\text{2nd}}$ $\boxed{\text{LIST}}$ $\boxed{\triangleright}$ 5 to paste "seq(" into the cumSum expression. Finally press $\boxed{\text{X, T, }\Theta, n}$ $\boxed{x^2}$ $\boxed{-}$ 3 $\boxed{,}$ $\boxed{\text{X, T, }\Theta, n}$ $\boxed{,}$ 1 $\boxed{,}$ 4 $\boxed{)}$ $\boxed{)}$ $\boxed{\text{ENTER}}$. We show the result with the calculator set in Seq mode. The variable will be expressed as x if the calculator is in Func mode.

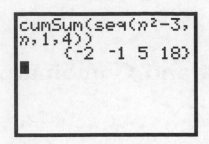

The TI-83 Plus and the TI-84 Plus can also compute individual partial sums. This is demonstrated in the next example.

Section 11.1, Example 7 (a) Evaluate $\displaystyle\sum_{k=1}^{5} k^3$.

Press $\boxed{\text{2nd}}$ $\boxed{\text{LIST}}$ $\boxed{\triangleright}$ $\boxed{\triangleright}$ 5 $\boxed{\text{2nd}}$ $\boxed{\text{LIST}}$ $\boxed{\triangleright}$ 5 $\boxed{\text{X, T, }\Theta, n}$ $\boxed{\wedge}$ 3 $\boxed{,}$ $\boxed{\text{X, T, }\Theta, n}$ $\boxed{,}$ 1 $\boxed{,}$ 5 $\boxed{)}$ $\boxed{)}$ $\boxed{\text{ENTER}}$. We show the result on a calculator set in Seq mode.

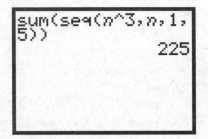

RECURSIVELY DEFINED SEQUENCES

Recursively defined sequences can also be entered on a graphing calculator set in Seq mode.

Section 11.1, Example 9 Find the first 5 terms of the sequence defined by

$$a_1 = 5, \; a_{k+1} = 2a_k - 3, \text{ for } k \geq 1.$$

Press $\boxed{\text{Y} =}$ and enter the recursive function beside "u(n) =" by pressing 2 $\boxed{\text{2nd}}$ $\boxed{\text{u}}$ $\boxed{(}$ $\boxed{\text{X,T,}\theta, n}$ $\boxed{-}$ 1 $\boxed{)}$ $\boxed{-}$ 3. (u is the second operation associated with the $\boxed{7}$ key.) Also set u(nMin) = 5, the first term of the sequence. Press $\boxed{\text{2nd}}$ $\boxed{\text{TblSet}}$ to display the TABLE SETUP screen and set TblStart = 1, ΔTbl = 1, and Indpnt: Auto. Now press $\boxed{\text{2nd}}$ $\boxed{\text{TABLE}}$ to display the table of values.

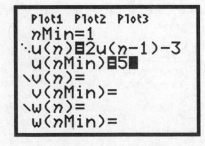 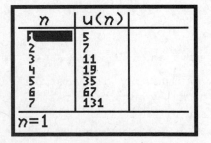

We see that $a_1 = 5$, $a_2 = 7$, $a_3 = 11$, $a_4 = 19$, and $a_5 = 35$.

EVALUATING FACTORIALS, PERMUTATIONS, AND COMBINATIONS

Operations from the MATH PRB (probability) menu can be used to evaluate factorials, permutations, and combinations.

Section 11.5, Exercise 6 Evaluate 7!.

Press 7 | MATH | ▷ | ▷ | ▷ | 4 | ENTER |. These keystrokes enter 7, display the MATH PRB menu, select item 4, !, from that menu, and then cause 7! to be evaluated. The result is 5040.

Section 11.5, Exercise 9 Evaluate $\frac{9!}{5!}$.

Press 9 | MATH | ▷ | ▷ | ▷ | 4 | ÷ | 5 | MATH | ▷ | ▷ | ▷ | 4 | ENTER |. The result is 3024.

```
7!
                5040
9!/5!
                3024
```

Section 11.5, Example 3 (a) Compute $_4P_4$.

Press 4 | MATH | ▷ | ▷ | ▷ | 2 4 | ENTER |. These keystrokes enter 4, for 4 objects, display the MATH PRB menu, select item 2, $_nP_r$, from that menu, enter 4, for 4 objects taken at a time, and then cause the calculation to be performed. The result is 24.

Section 11.5, Example 6 Compute $_8P_4$.

Press 8 | MATH | ▷ | ▷ | ▷ | 2 4 | ENTER |. These keystrokes enter 8, for 8 objects, display the MATH PRB menu, select item 2, $_nP_r$, from that menu, enter 4, for 4 objects taken at a time, and then cause the calculation to be performed. The result is 1680.

```
4 nPr 4
                  24
8 nPr 4
                1680
■
```

Section 11.6, Example 2 Evaluate $\begin{pmatrix} 7 \\ 5 \end{pmatrix}$.

Press 7 | MATH | | ▷ | | ▷ | | ▷ | 3 5 | ENTER |. These keystrokes enter 7, for 7 objects, display the MATH PRB menu, select item 3, $_nC_r$, from that menu, enter 5, for 5 objects taken at a time, and then cause the calculation to be performed. The result is 21.

```
7 nCr 5
                    21
■
```

The TI-89 Graphing Calculator

Chapter R
Basic Concepts of Algebra

GETTING STARTED

Before turning on the calculator note that there are options above the keys as well as on them. To access the option written on a key, simply press the key. The options written in yellow above the keys are accessed by first pressing the yellow $\boxed{\text{2nd}}$ key in the left column of the keypad and then pressing the key corresponding to the desired option. The purple options above the keys are accessed by first pressing the purple $\boxed{\text{alpha}}$ key in the second column of the keypad. The green options are accessed by first pressing the $\boxed{\diamond}$ key in the left column of the keypad. This key has a green diamond inside a green border.

Press $\boxed{\text{ON}}$ to turn on the TI-89 graphing calculator. ($\boxed{\text{ON}}$ is the key at the bottom left-hand corner of the keypad.) The home screen is displayed. You should see a row of boxes at the top of the screen and two horizontal lines with lettering below them at the bottom of the screen. If you do not see anything, try adjusting the display contrast. To do this, first press and hold down the $\boxed{\diamond}$ key. Then press and hold $\boxed{+}$ to darken the display or $\boxed{-}$ to lighten the display. Be sure to use the black $\boxed{-}$ key in the right column of the keypad rather than the gray $\boxed{(-)}$ key on the bottom row.

One way to turn the calculator off is to press $\boxed{\text{2nd}}$ $\boxed{\text{OFF}}$. (OFF is the second operation associated with the $\boxed{\text{ON}}$ key.) When you turn the TI-89 on again the home screen will be displayed regardless of the screen that was displayed when the calculator was turned off. $\boxed{\text{2nd}}$ $\boxed{\text{OFF}}$ cannot be used to turn off the calculator if an error message is displayed. The calculator can also be turned off by pressing $\boxed{\diamond}$ $\boxed{\text{OFF}}$. This will work even if an error message is displayed. After $\boxed{\diamond}$ $\boxed{\text{OFF}}$ is used, when the TI-89 is turned on again the display will be exactly as it was when it was turned off. The calculator will turn itself off automatically after several minutes without any activity. When this happens the display will be just as you left it when you turn the calculator on again.

From top to bottom, the home screen consists of the toolbar, the large history area where entries and their corresponding results are displayed, the entry line where expressions or instructions are entered, and the status line which shows the current state of the calculator, These areas will be discussed in more detail as the need arises.

Press $\boxed{\text{MODE}}$ to display the MODE settings. Modes that are not currently valid, due to the existing choices of settings, are dimmed. Initially you should select the settings shown below.

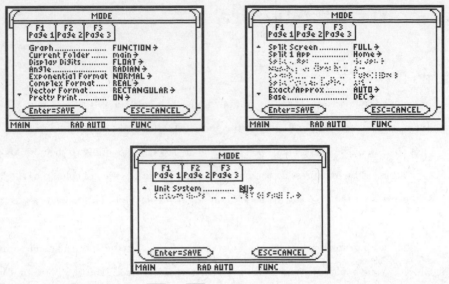

To change a setting on the Mode screen use $\boxed{\triangledown}$ or $\boxed{\triangle}$ to move the cursor to the line of that setting. Then use $\boxed{\triangleright}$ to display the options. Press the number of the desired option to copy it to the Mode screen. Then press $\boxed{\text{ENTER}}$ to save it there. Instead of pressing the number of the desired setting, you can highlight it and then press $\boxed{\text{ENTER}}$ $\boxed{\text{ENTER}}$ to copy it to the Mode screen and save it there. Note that the cursor skips dimmed settings as you move through the options.

It will be helpful to read Chapter 1: Getting Started and Chapter 2: Operating the TI-89 in the TI-89 Guidebook before proceeding.

USING A MENU

A menu is a list of options that appear when a key is pressed. For example, press $\boxed{\text{F1}}$ to display the Tools menu. We can select an item from a menu by using $\boxed{\triangledown}$ to highlight it and then pressing $\boxed{\text{ENTER}}$ or by simply pressing the number of the item. If we press 8 when the Tools menu is displayed, for instance, we select "Clear Home" and any previously entered computations will be cleared from the history area of the home screen. If an item is identified by a letter rather than a number, press the purple $\boxed{\text{alpha}}$ key followed by the letter of the item to select it. The letters are printed in purple above the keys on the keypad. The down-arrow beside item 8 in the menu below indicates that there are additional items in the menu. Use $\boxed{\triangledown}$ to scroll down to them.

ABSOLUTE VALUE

Section R.1, Example 3 Find the distance between -2 and 3.

First press $\boxed{\text{HOME}}$ or $\boxed{\text{2nd}}$ $\boxed{\text{QUIT}}$ to go to the home screen. You might want to clear any previously entered computations from the history area of the home screen first. To do this, access Tools from the toolbar at the top of the screen by pressing $\boxed{\text{F1}}$, the blue key at the top left-hand corner of the keypad. Then select item 8, Clear Home, from this menu by pressing 8.

The entry line on the home screen can be cleared by pressing $\boxed{\text{CLEAR}}$. This is not necessary if the current entry is highlighted, since it will automatically be cleared when the first character of a new entry is entered.

The distance between -2 and 3 is $|-2-3|$, or $|3-(-2)|$. Absolute value notation is denoted "abs" on the TI-89. It is item 2 on the MATH Number menu. Note that the gray $\boxed{(-)}$ key in the bottom row of the keypad must be used to enter a negative number. The black $\boxed{-}$ key in the right column of the keypad is used for the subtraction operation.

To enter $|-2-3|$ press $\boxed{\text{2nd}}$ $\boxed{\text{MATH}}$ $\boxed{\triangleright}$ 2 $\boxed{(-)}$ 2 $\boxed{-}$ 3 $)$ $\boxed{\text{ENTER}}$. (MATH is the second operation associated with the 5 numeric key.)To enter $|3-(-2)|$ press $\boxed{\text{2nd}}$ $\boxed{\text{MATH}}$ $\boxed{\triangleright}$ 2 3 $\boxed{-}$ $\boxed{(}$ $\boxed{(-)}$ 2 $\boxed{)}$ $\boxed{)}$ $\boxed{\text{ENTER}}$. Note that the calculator supplies the left parenthesis in the absolute value notation. We close the expression with a right parenthesis. Although the parentheses around -2 in the second expression are not necessary, they allow the expression to be more easily read on the entry line so we include them here.

Instead of pressing $\boxed{\text{2nd}}$ $\boxed{\text{MATH}}$ $\boxed{\triangleright}$ 2 to access "abs(" and copy it to the home screen, we could have pressed $\boxed{\text{2nd}}$ $\boxed{\text{MATH}}$ $\boxed{\triangleright}$ $\boxed{\triangledown}$ $\boxed{\text{ENTER}}$. Absolute value notation can also be found as the first item in the Catalog and copied to the home screen. Access the Catalog by pressing $\boxed{\text{CATALOG}}$. Then use the $\boxed{\triangle}$ or $\boxed{\triangledown}$ key to position the triangular selection cursor beside "abs(." This cursor can be positioned quickly by pressing $\boxed{\text{A}}$ to move the cursor to the first item in the Catalog that begins with A, "abs(." (A is the purple alphabetic operation associated with the $\boxed{=}$ key.) Note that, when we are in the Catalog, it is not necessary to press $\boxed{\text{alpha}}$ before an alphabetic entry.

EDITING ENTRIES

You can edit your entry if necessary. After ENTER is pressed to evaluate an expression, the TI-89 leaves the expression on the entry line and highlights it. To edit the expression you must first remove the highlight to avoid the possibility of accidently typing over the entire expression. To do this, press ◁ or ▷ to move the cursor (a blinking vertical line) toward the side of the expression to be edited. If, for instance, in entering one of the expressions in Example 3 above you pressed 6 instead of 3, first press ◁ to move the cursor to the beginning of the expression or ▷ to move it to the end of the expression. Now, to type a 6 over the 3, first select overtype mode by pressing 2nd INS . (INS is the second operation associated with the ← key.) Now the cursor becomes a dark, blinking rectangle rather than a vertical line. Use ▷ to position the cursor over the 6 and then press 3. To leave overtype mode press 2nd INS again. The calculator is now in the insert mode, indicated by a vertical cursor, and will remain in that mode until overtype mode is once again selected.

If you forgot to type the 2 in the first expression, move the insert cursor to the left of the subtraction symbol and press 2 to insert the 2 before that symbol. You can continue to insert symbols immediately after the first insertion. If you typed 21 instead of 2, move the cursor to the left of 1 and press ← . This will delete the 1. Instead of using overtype mode to overtype a character as described above, we can use ← to delete the character and then, in insert mode, insert a new character.

If you accidently press △ instead of ◁ or ▷ while editing an expression, the cursor will move up into the history area of the screen. Press ESC to return immediately to the entry line. The ▽ key can also be used to return to the entry line. It must be pressed the same number of times the △ key was pressed accidently.

If you notice that an entry needs to be edited before you press ENTER to perform the computation, the editing can be done as described above without the necessity of first removing the highlight from the entry.

The keystrokes 2nd ENTRY can be used repeatedly to recall entries preceding the last one. (ENTRY is the second function associated with the ENTER key.) Pressing 2nd ENTRY twice, for example, will recall the next to last entry. Using these keystrokes a third time recalls the third to last entry and so on. The number of entries that can be recalled depends on the amount of storage they occupy in the calculator's memory.

Previous entries and results of computations can also be copied to the entry line by first using the △ key to move through the history area until the desired entry or result is highlighted. Then press ENTER to copy it to the entry line.

SCIENTIFIC NOTATION

To enter a number in scientific notation, first type the decimal portion of the number; then press the EE key in the left column of the keypad; finally type the exponent, which can be at most three digits. For example, to enter 1.789×10^{-11} in scientific notation, go to the home screen and press 1 . 7 8 9 EE (−) 1 1 ENTER . To enter 6.084×10^{23} in scientific notation, press 6 . 0 8 4 EE 2 3 ENTER . The decimal portion of each number appears before a small E while the exponent follows the E.

```
┌────────────────────────────────────────┐
│ ⌐F1┐⌐F2┐⌐F3┐⌐F4┐┌F5 ┐⌐F6┐              │
│ Tools A19ebra Calc Other Pr9miO Clean Up│
│                                          │
│                                          │
│  ■ 1.789ᴇ⁻11          1.789ᴇ⁻11         │
│  ■ 6.084ᴇ23           6.084ᴇ23          │
│ ┌6.084ᴇ23─────────────────────────────┐ │
│ MAIN        RAD AUTO    FUNC     2/30   │
└────────────────────────────────────────┘
```

The TI-89 can be used to perform computations in scientific notation.

Section R.2, Example 8 *Distance to a Star.* The nearest star, Alpha Centauri C, is about 4.22 light-years from Earth. One light-year is the distance that light travels in one year and is about 5.88×10^{12} miles. How many miles is it from Earth to Alpha Centauri C? Express your answer in scientific notation.

To solve this problem we find the product $4.22 \times (5.88 \times 10^{12})$. Press 4 $\boxed{.}$ 2 2 $\boxed{\times}$ 5 $\boxed{.}$ 8 8 $\boxed{\text{EE}}$ 1 2 $\boxed{\text{ENTER}}$. The result is 2.48136×10^{13} miles.

```
┌────────────────────────────────────────┐
│ ⌐F1┐⌐F2┐⌐F3┐⌐F4┐┌F5 ┐⌐F6┐              │
│ Tools A19ebra Calc Other Pr9miO Clean Up│
│                                          │
│                                          │
│  ■ 4.22·5880000000000.                   │
│                        2.48136ᴇ13        │
│ ┌4.22*5.88ᴇ12──────────────────────────┐│
│ MAIN        DEG AUTO    FUNC            ││
└────────────────────────────────────────┘
```

ORDER OF OPERATIONS

Section R.2, Example 9 (b) Calculate: $\dfrac{10 \div (8 - 6) + 9 \cdot 4}{2^5 + 3^2}$.

In order to divide the entire numerator by the entire denominator, we must enclose both the numerator and the denominator in parentheses. That is, we enter $(10 \div (8 - 6) + 9 \cdot 4) \div (2^5 + 3^2)$. On the home screen press $\boxed{(}$ 1 0 $\boxed{\div}$ $\boxed{(}$ 8 $\boxed{-}$ 6 $\boxed{)}$ $\boxed{+}$ 9 $\boxed{\times}$ 4 $\boxed{)}$ $\boxed{\div}$ $\boxed{(}$ 2 $\boxed{\wedge}$ 5 $\boxed{+}$ 3 $\boxed{\wedge}$ 2 $\boxed{)}$ $\boxed{\text{ENTER}}$.

```
┌────────────────────────────────────────┐
│ ⌐F1┐⌐F2┐⌐F3┐⌐F4┐┌F5 ┐⌐F6┐              │
│ Tools A19ebra Calc Other Pr9miO Clean Up│
│                                          │
│          10                              │
│        ─────── + 9·4                     │
│   ■     8 - 6                      1     │
│        ─────────────                     │
│          2⁵ + 3²                         │
│ ┌(10/(8-6)+9*4)/(2^5+3^2)──────────────┐ │
│ MAIN        RAD AUTO    FUNC     1/30   │
└────────────────────────────────────────┘
```

RADICAL NOTATION

We can use the square-root key and rational exponents to simplify radical expressions.

Section R.7, Example 1 Simplify each of the following.

a) $\sqrt{36}$ **b)** $-\sqrt{36}$ **c)** $\sqrt[3]{-8}$ **d)** $\sqrt[5]{\dfrac{32}{243}}$ **e)** $\sqrt[4]{-16}$

a) To find $\sqrt{36}$ go to the home screen and press $\boxed{\text{2nd}}$ $\boxed{\sqrt{}}$ 3 6 $\boxed{)}$ $\boxed{\text{ENTER}}$. ($\sqrt{}$ is the second operation associated with the $\boxed{\times}$ multiplication key.) Note that the calculator supplies a left parenthesis with the radical symbol and we must close the expression with a right parenthesis.

b) To find $-\sqrt{36}$ press $\boxed{(-)}$ $\boxed{\text{2nd}}$ $\boxed{\sqrt{}}$ 3 6 $\boxed{)}$ $\boxed{\text{ENTER}}$.

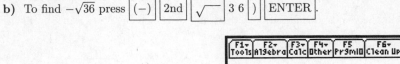

c) We will also use a rational exponent to find $\sqrt[3]{-8}$. We have $\sqrt[3]{-8} = (-8)^{1/3}$. Press $\boxed{(}$ $\boxed{(-)}$ 8 $\boxed{)}$ $\boxed{\wedge}$ $\boxed{(}$ 1 $\boxed{\div}$ 3 $\boxed{)}$ $\boxed{\text{ENTER}}$.

d) We will use a rational exponent to find $\sqrt[5]{\dfrac{32}{243}}$. We have $\sqrt[5]{\dfrac{32}{243}} = \left(\dfrac{32}{243}\right)^{1/5}$. Press $\boxed{(}$ 3 2 $\boxed{\div}$ 2 4 3 $\boxed{)}$ $\boxed{\wedge}$ $\boxed{(}$ 1 $\boxed{\div}$ 5 $\boxed{)}$ $\boxed{\text{ENTER}}$.

e) We have $\sqrt[4]{-16} = (-16)^{1/4}$. Press $\boxed{(}$ $\boxed{(-)}$ 1 6 $\boxed{)}$ $\boxed{\wedge}$ $\boxed{(}$ 1 $\boxed{\div}$ 4 $\boxed{)}$ $\boxed{\text{ENTER}}$. When Real is selected for Complex Format on the MODE screen, the calculator returns an error message indicating that the result is not a real number.

Chapter 1
Graphs, Functions, and Models

GRAPHING EQUATIONS

Section 1.1, Example 4 Graph $3x - 5y = -10$.

First we solve for y as shown in the text, obtaining $y = \dfrac{3}{5}x + 2$. Then press ◊ $\boxed{Y =}$ to access the equation-editor screen. (Y = is the green ◊ operation associated with the $\boxed{F1}$ key.) If any plots are turned on they should be turned off, or deselected, now. A check mark beside the name of a plot indicates that it is currently selected. To deselect it, move the cursor to the plot. Then press $\boxed{F4}$. There should now be no check mark beside the name of the plot, indicating that it has been deselected. If there is currently an expression displayed for y_1, clear it by positioning the cursor beside "y1 =" and then press $\boxed{\text{CLEAR}}$. Do the same for expressions that appear on all other "y =" lines by using $\boxed{\triangledown}$ to move to a line and then pressing $\boxed{\text{CLEAR}}$. Then use $\boxed{\triangle}$ or $\boxed{\triangledown}$ to move the cursor beside "y1 =." Now enter $y_1 = \dfrac{3}{5}x + 2$ on the entry line of the equation-editor screen and paste it beside "y1 =" by pressing $\boxed{(}$ $\boxed{3}$ $\boxed{\div}$ $\boxed{5}$ $\boxed{)}$ $\boxed{X}$ $\boxed{+}$ 2 $\boxed{\text{ENTER}}$. Although the parentheses are not necessary, the equation is more easily read on the entry line when they are used.

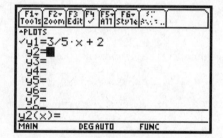

Next we set the viewing window. This is the portion of the coordinate plane that appears on the calculator's screen. It is defined by the minimum and maximum values of x and y: xmin, xmax, ymin, and ymax. The notation [xmin, xmax, ymin, ymax] is used in the text to represent these window settings or dimensions. For example, $[-12, 12, -8, 8]$ denotes a window that displays the portion of the x-axis from -12 to 12 and the portion of the y-axis from -8 to 8. In addition, the distance between tick marks on the axes is defined by the settings xscl and yscl. In this manual xscl and yscl will be assumed to be 1 unless noted otherwise. The setting xres sets the pixel resolution. We usually select xres = 2. The window corresponding to the settings $[-20, 30, -12, 20]$, xscl = 5, yscl = 2, xres = 2, is shown below.

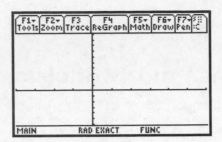

Press $\boxed{\diamond}$ $\boxed{\text{WINDOW}}$ to display the current window settings on your calculator. (WINDOW is the green $\diamond$ operation associated with the $\boxed{\text{F2}}$ key on the top row of the keypad.) The standard settings are shown below.

To change a setting, position the cursor beside the setting you wish to change and enter the new value. For example, to change from the standard settings to $[-20, 30, -12, 20]$, xscl $= 5$, yscl $= 2$, on the WINDOW screen, start with the setting beside "xmin =" highlighted and press $\boxed{(-)}$ 2 0 $\boxed{\text{ENTER}}$ 3 0 $\boxed{\text{ENTER}}$ 5 $\boxed{\text{ENTER}}$ $\boxed{(-)}$ 1 2 $\boxed{\text{ENTER}}$ 2 0 $\boxed{\text{ENTER}}$ 2 $\boxed{\text{ENTER}}$. You must use the gray $\boxed{(-)}$ key on the bottom row of the keypad rather than the black $\boxed{-}$ key in the right-hand column to enter a negative number. $\boxed{(-)}$ represents "the opposite of" or "the additive inverse of" whereas $\boxed{-}$ is the key for the subtraction operation. The $\boxed{\nabla}$ key may be used instead of $\boxed{\text{ENTER}}$ after typing each window setting. To see the window $[-20, 30, -12, 20]$, xscl $= 5$, yscl $= 2$ shown above, press $\boxed{\diamond}$ $\boxed{\text{GRAPH}}$. (GRAPH is the green $\diamond$ operation associated with the $\boxed{\text{F3}}$ key on the top row of the keypad.)

QUICK TIP: To return quickly to the standard window setting $[-10, 10, -10, 10]$, xscl $= 1$, yscl $= 1$, when either the Window screen or the Graph screen is displayed, press $\boxed{\text{F2}}$ to access the ZOOM menu and then press 6 to select item 6, ZoomStd (Zoom Standard).

The standard window is a good choice for the graph of the equation $y = \frac{3}{5}x + 2$. Either enter these dimensions in the WINDOW screen and then press $\boxed{\diamond}$ $\boxed{\text{GRAPH}}$ to see the graph or, from the WINDOW screen, simply press $\boxed{\text{F2}}$ 6 to select the standard window and see the graph.

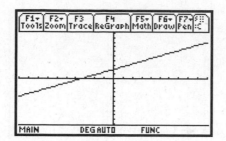

THE TABLE FEATURE

A table of x-and y-values representing ordered pairs that are solutions of an equation can be displayed on the TI-89.

Section 1.1, Example 5 Create a table of ordered pairs that are solutions of the equation $y = x^2 - 9x - 12$.

First enter the equation on the equation-editor screen. Once the equation in entered, press $\diamond$ $\boxed{\text{TblSet}}$ or $\diamond$ $\boxed{\text{TABLE}}$ $\boxed{\text{F2}}$ to access the TABLE SETUP window. (TblSet is the green $\diamond$ operation associated with the $\boxed{\text{F4}}$ key.) If "Independent" is set to "Auto" on the Table Setup screen, the calculator will supply values for x, beginning with the value specified as tblStart and continuing by adding the value of Δtbl to the preceding value for x. If the table was previously set to Ask, the blinking cursor will be positioned over ASK. Change this setting to AUTO by pressing $\boxed{\triangleright}$ 1. Now use the $\boxed{\triangle}$ key to move the cursor to tblStart. Enter a minimum x-value of -3, an increment of 1, and a Graph $<->$ Table setting of OFF by first positioning the cursor beside tblStart and then pressing $\boxed{(-)}$ 3 $\boxed{\triangledown}$ 1 $\boxed{\triangledown}$ $\boxed{\triangleright}$ 1 $\boxed{\text{ENTER}}$. Press $\diamond$ $\boxed{\text{TABLE}}$ to see the table. (TABLE is the green $\diamond$ operation associated with the $\boxed{\text{F5}}$ key.) We can use the $\boxed{\triangle}$ and $\boxed{\triangledown}$ keys to scroll up and down in the table to find values other than those shown here.

SQUARING THE VIEWING WINDOW

In the standard window, the distance between tick marks on the y-axis is about 1/2 the distance between tick marks on the x-axis. It is often desirable to choose window dimensions for which these distances are the same, creating a "square" window. Any window in which the ratio of the length of the y-axis to the length of the x-axis is 1/2 will produce this effect.

This can be accomplished by selecting dimensions for which ymax $-$ ymin $= \dfrac{1}{2}$(xmax $-$ xmin). For example, the windows $[-12, 12, -6, 6]$ and $[-6, 6, -3, 3]$ are square. To illustrate this, we graph the circle $x^2 + y^2 = 9$ in the standard window. Note that the graph does not appear to be a circle. (We will explain how to graph a circle later in this chapter.)

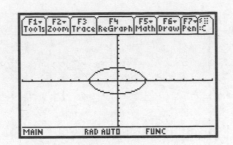

Now change the window dimensions to $[-8, 8, -4, 4]$, xscl $= 1$, yscl $= 1$, and press $\boxed{\diamond}$ $\boxed{\text{GRAPH}}$. Observe that the distance between tick marks appears to be the same on both axes and that the graph appears to be a circle.

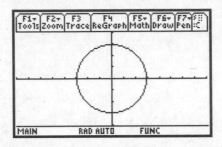

The window can also be squared using the calculator's ZoomSqr feature. From the equation-editor, Window, or Graph screen, press $\boxed{\text{F2}}$ 5 to select the ZoomSqr window. The resulting window dimensions and graph are shown below. Here the window was squared from the standard window. Note that the graph also appears to be a circle in this window.

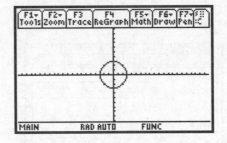

GRAPHING CIRCLES

If the center and radius of a circle are known, the circle can be graphed using the Circle feature from the Catalog.

Section 1.1, Example 12 Graph $(x - 2)^2 + (y + 1)^2 = 16$.

The center of this circle is $(2, -1)$ and its radius is 4. To graph it first press $\boxed{\diamond}$ $\boxed{\text{Y} =}$ and clear all previously entered equations on the equation-editor screen. Then select a square window. The dimensions $[-12, 12, -6, 6]$ are a good choice for this circle. Now, to select Circle from the Catalog, first press $\boxed{\text{HOME}}$ or $\boxed{\text{2nd}}$ $\boxed{\text{QUIT}}$ to go to the home screen. Then press $\boxed{\text{CATALOG}}$ $\boxed{\text{C}}$ to go to the beginning of the items in the Catalog that start with C. (Note that it is not necessary to press $\boxed{\text{alpha}}$ before a letter key when the Catalog is displayed.) Now use $\boxed{\triangledown}$ to go to "Circle" and press $\boxed{\text{ENTER}}$. The Circle command appears on the entry line of the home screen. "Circle" can also be typed directly on the entry line of the home screen by pressing $\boxed{\text{2nd}}$ $\boxed{\text{a-lock}}$ $\boxed{\text{C}}$ $\boxed{\text{I}}$ $\boxed{\text{R}}$ $\boxed{\text{C}}$ $\boxed{\text{L}}$ $\boxed{\text{E}}$ $\boxed{\text{alpha}}$. Enter the x-coordinate of the center, the y-coordinate of the center, and the length of the radius, all separated by commas. To do this press 2 $\boxed{,}$ $\boxed{(-)}$ 1 $\boxed{,}$ 4. Press $\boxed{\text{ENTER}}$ to see the graph.

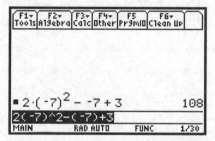

This graph can be cleared from the Graph screen by pressing $\boxed{\text{F4}}$ (ReGraph) or by pressing $\boxed{\text{2nd}}$ $\boxed{\text{F6}}$ to display the Draw menu and then pressing 1 to select ClrDraw. The ClrDraw command can also be accessed from the Catalog. From the home screen, press $\boxed{\text{CATALOG}}$ $\boxed{\text{C}}$, scroll to ClrDraw, and press $\boxed{\text{ENTER}}$ $\boxed{\text{ENTER}}$.

FINDING FUNCTION VALUES

When a formula for a function is given, function values can be found in several ways.

Section 1.2, Example 4 (b) For $f(x) = 2x^2 - x + 3$, find $f(-7)$.

Method 1: Substitute the inputs directly in the formula. On the home screen press 2 $\boxed{(}$ $\boxed{(-)}$ 7 $\boxed{)}$ $\boxed{\wedge}$ 2 $\boxed{-}$ $\boxed{(}$ $\boxed{(-)}$ 7 $\boxed{)}$ $\boxed{+}$ 3 $\boxed{\text{ENTER}}$.

Method 2: Enter $y_1 = 2x^2 - x + 3$ on the "y =" screen. Then press $\boxed{\text{HOME}}$ or $\boxed{\text{2nd}}$ $\boxed{\text{QUIT}}$ to go to the home screen. To find $f(-7)$, the value of y_1 when $x = -7$, press $\boxed{(-)}$ 7 $\boxed{\text{STO} \triangleright}$ $\boxed{\text{X}}$ $\boxed{\text{2nd}}$ $\boxed{:}$ $\boxed{\text{Y}}$ 1 $\boxed{(}$ $\boxed{\text{X}}$ $\boxed{)}$ $\boxed{\text{ENTER}}$. (: is the second operation associated with the 4 numeric key.) This series of keystrokes stores -7 as the value of x and then substitutes it in the function y_1.

Method 3: Enter $y_1 = 2x^2 - x + 3$ on the "Y =" screen and press $\boxed{\text{HOME}}$ or $\boxed{\text{2nd}}$ $\boxed{\text{QUIT}}$ to go to the home screen. Then press $\boxed{\text{Y}}$ 1 $\boxed{(}$ $\boxed{(-)}$ 7 $\boxed{)}$ $\boxed{\text{ENTER}}$. Note that this entry closely resembles function notation.

```
┌─────────────────────────────────────────┐
│ F1┬ F2┬  F3┬ F4┬  F5    F6┬              │
│Tools Algebra Calc Other PrgmIO Clean Up  │
│                                          │
│                                          │
│                                          │
│ ■ y1( -7)                          108   │
│ y1(-7)                                   │
│MAIN        RAD AUTO      FUNC      1/30   │
└─────────────────────────────────────────┘
```

Method 4: The TABLE feature can also be used to find function values. Enter $y_1 = 2x^2 - x + 3$ on the "Y =" screen. Then set up a table in Ask mode. Press $\boxed{\diamond}$ $\boxed{\text{TblSet}}$ or $\boxed{\diamond}$ $\boxed{\text{TABLE}}$ $\boxed{\text{F2}}$ to access the TableSetup screen. Move the cursor to the "Independent" line. Then press $\boxed{\triangleright}$ 2 $\boxed{\text{ENTER}}$ to select Ask mode. In Ask mode the calculator disregards the other settings on the Table Setup screen.

```
┌─────────────────────────────────────────┐
│ F1┬  F2┬  F3  F4  F5┬ F6┬                │
│              TABLE SETUP                  │
│                                          │
│  tblStart:         0.                    │
│  Δtbl:             0.                    │
│  Graph <-> Table:  OFF                   │
│  Independent:      AUTO→                 │
│ ⟨Enter=SAVE⟩        ⟨ESC=CANCEL⟩          │
│ y2(x)=                                    │
│MAIN        RAD EXACT     FUNC            │
└─────────────────────────────────────────┘
```

Now press $\boxed{\diamond}$ $\boxed{\text{TABLE}}$ to view the table. (TABLE is the second operation associated with the $\boxed{\text{F5}}$ key.) If you select Ask before a table is displayed for the first time on your calculator, a blank table is displayed. If a table has previously been displayed, the table you now see will continue to show the values in the previous table.

Values for x can be entered in the x-column of the table and the corresponding values for y_1 will be displayed in the $y1$-column. To enter -7, press $\boxed{(-)}$ 7 $\boxed{\text{ENTER}}$. Any additional x-values that are displayed are from a table that was previously displayed on the Auto setting. We see that $y_1 = 108$ when $x = -7$, so $f(-7) = 108$.

```
┌─────────────────────────────────────────┐
│ F1┬  F2    F3   F4     F5      F6        │
│Tools Setup Cell Header Del Row Ins Row   │
│ x      y1                                │
│ -7.    108.                              │
│ ▓▓▓▓▓                                    │
│                                          │
│                                          │
│                                          │
│ x=                                       │
│MAIN        RAD AUTO      FUNC            │
└─────────────────────────────────────────┘
```

Method 5: We can also use the Value feature from the Math menu on the GRAPH screen to find $f(-7)$. To do this, graph $y_1 = 2x^2 - x + 3$ in a window that includes the x-value -7. We will use the standard window. Then press $\boxed{\text{F5}}$ $\boxed{\text{ENTER}}$ to access the Math menu and select item 1, Value. Now supply the desired x-value by pressing $\boxed{(-)}$ 7. Press $\boxed{\text{ENTER}}$ to see $x = -7$, $y = 108$ at the bottom of the screen, Thus, $f(-7) = 108$.

GRAPHS OF FUNCTIONS

Three functions are graphed in **Section 1.2, Example 5**. The TI-89 does not use function notation. Consequently, we must first replace function notation with y when we graph a function on this calculator. For example, to graph $f(x) = x^2 - 5$ replace $f(x)$ with y. Then enter the equation $y = x^2 - 5$ on the equation-editor screen and graph it as described on page 85 of this manual.

LINEAR REGRESSION

We can use the Linear Regression feature to fit a linear equation to a set of data.

Section 1.4, Example 7 The following table shows the number of U. S. households subscribing to cable television, in millions, for the years 1999 through 2006. Fit a regression line to the data using the linear regression feature on a graphing calculator. Then use the linear model to estimate the number of cable television subscribers in 2010.

Years, x	U. S. Households with Cable Television (in millions)
1999, 0	76.4
2000, 1	78.6
2001, 2	81.5
2002, 3	87.8
2003, 4	88.4
2004, 5	92.4
2005, 6	94.0
2006, 7	95.0

We will enter the data as ordered pairs in the Data/Matrix editor. Press $\boxed{\text{APPS}}$ 6 3 to display a new data variable screen in the Data/Matrix editor. We must now enter a data variable name in the Variable box on this screen. The name can contain from 1 to 8 characters and cannot start with a numeral. Some names are preassigned to other uses on the TI-89. If you try to use one of these, you will get an error message. Press $\boxed{\triangledown}$ $\boxed{\triangledown}$ to move the cursor to the Variable box. We will name our data variable "cable." To enter this name, first lock the alphabetic keys on by pressing $\boxed{\text{2nd}}$ $\boxed{\text{a-lock}}$. Then press $\boxed{\text{C}}$ $\boxed{\text{A}}$ $\boxed{\text{B}}$ $\boxed{\text{L}}$ $\boxed{\text{E}}$. The letters C, A, B, L, and E are the purple alphabetic operations associated with the $\boxed{)}$, $\boxed{=}$, $\boxed{(}$, 4, and $\boxed{\div}$ keys, respectively.

```
┌─────────────────────────────────────────┐
│ ╔═══════════════════════════════════════╗ │
│ ║              NEW                       ║ │
│ ║  Type:        Data →                   ║ │
│ ║  Folder:      main →                   ║ │
│ ║  Variable:   [cable        ]           ║ │
│ ║  ...........: [            ]            ║ │
│ ║  ...........: [            ]            ║ │
│ ║                                        ║ │
│ ║  ⟨ Enter=OK ⟩        ⟨ ESC=CANCEL ⟩     ║ │
│ ╚═══════════════════════════════════════╝ │
│  MAIN      ▮▮▮ DEG AUTO    FUNC            │
└─────────────────────────────────────────┘
```

After typing the name of the data variable, unlock the alphabetic keys by pressing the purple [alpha] key. Now press [ENTER] [ENTER] to go to the data-entry screen. Assuming the data variable name "cable" has not previously been used in your calculator, this screen will contain empty data lists with row 1, column 1 highlighted. If entries have previously been made in a data variable named "cable," they can be cleared by pressing [F1] 8 [ENTER].

We will enter the first coordinates of the points as the number of years since 1999 in column c1 and the second coordinates in c2. To enter 0 press 0 [ENTER]. Continue typing the first coordinates 1, 2, 3, 4, 5, 6, and 7, each followed by [ENTER]. The entries can be followed by [▽] rather than [ENTER] if desired. Press [▷] [△] [△] [△] [△] [△] [△] to move to the top of column c2. Type the second coordinates 76.4, 78.6, 81.5, 87.8, 88.4, 92.4, 94.0, and 95.0 in succession, each followed by [ENTER] or [▽]. Note that the coordinates of each point must be in the same position in both lists.

```
┌─────────────────────────────────────┐   ┌─────────────────────────────────────┐
│ F1▾│ F2      │ F3 │ F4    │ F5 │F6▾│F7│   │ F1▾│ F2      │ F3 │ F4    │ F5 │F6▾│F7│
│Tools│Plot Setup│Cell│Header│Calc│Util│Stat│ │Tools│Plot Setup│Cell│Header│Calc│Util│Stat│
│ DATA │          │       │       │       │   │ DATA │          │       │       │       │
│      │ c1       │ c2    │ c3    │       │   │      │ c1       │ c2    │ c3    │       │
│ 1    │ 0        │ 76.4  │ ▮▮▮▮▮ │       │   │ 5    │ 4        │ 88.4  │       │       │
│ 2    │ 1        │ 78.6  │       │       │   │ 6    │ 5        │ 92.4  │       │       │
│ 3    │ 2        │ 81.5  │       │       │   │ 7    │ 6        │ 94    │       │       │
│ 4    │ 3        │ 87.8  │       │       │   │ 8    │ 7        │ 95    │ ▮▮▮▮▮ │       │
│ r1c3=                                  │   │ r8c3=                                  │
│ MAIN      DEG AUTO      FUNC           │   │ MAIN      DEG AUTO      FUNC           │
└─────────────────────────────────────┘   └─────────────────────────────────────┘
```

The calculator can then plot these points, creating a scatterplot of the data. To do this we first access the Plot Setup screen by pressing [F2]. We will use Plot 1, which is highlighted. If any plot settings are currently entered beside "Plot 1," clear them by pressing [F3]. Clear settings shown beside any other plots as well by using [▽] to highlight each plot in turn and then pressing [F3].

Now we define Plot 1. Use [△] to highlight Plot 1 if necessary. Then press [F1] to display the Plot Definition screen. The item on the first line, Plot Type, is highlighted. We will choose a scatterplot, by pressing [▷] 1. Now press [▽] to go to the next line, Mark. Here we select the type of mark or symbol that will be used to plot the points. We select a box by pressing [▷] 1. Now we must tell the calculator which columns of the data variable to use for the x- and y-coordinates of the points to be plotted. Press [▽] to move the cursor to the "x" line and enter c1 as the source of the x-coordinates by pressing [alpha] [C] 1. (C is the purple alphabetic operation associated with the [)] key.) Press [▽] [alpha] [C] 2 to go the the "y" line and enter c2 as the source of the y-coordinates.

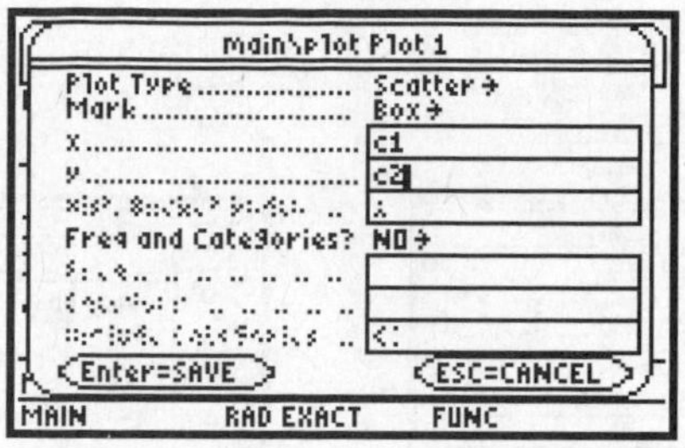

Save the plot definition and return to the Plot Setup screen by pressing $\boxed{\text{ENTER}}$ $\boxed{\text{ENTER}}$. Beside "Plot 1:" you will now see a shorthand notation for the definition of the plot. The check mark to the left of Plot 1 indicates that it is turned on.

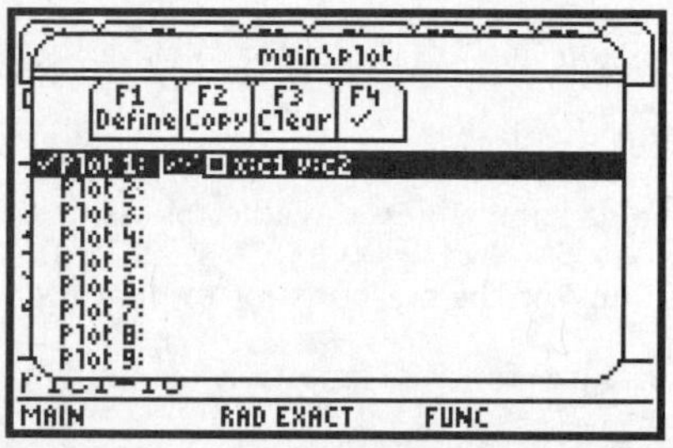

Note that there should be no equations entered on the equation-editor screen. If there are equations present, clear them as described on page 85 of this manual.

Now we select a window. The x-values range from 0 through 7 and the y-values from 76.4 through 95.0, so one good choice is $[-1, 8, 70, 100]$, xscl $= 1$, yscl $= 5$. We can enter these dimensions and then press $\boxed{\diamond}$ $\boxed{\text{GRAPH}}$ to see the scatterplot. Instead of entering the window dimensions directly, we can press $\boxed{\text{F2}}$ 9 after entering the coordinates of the points in lists and defining Plot 1. This activates the ZoomData operation which automatically defines a viewing window that displays all the points and also displays the graph. We use the window $[-1, 8, 70, 100]$, xscl $= 1$, yscl $= 5$ to produce the scatterplot shown below.

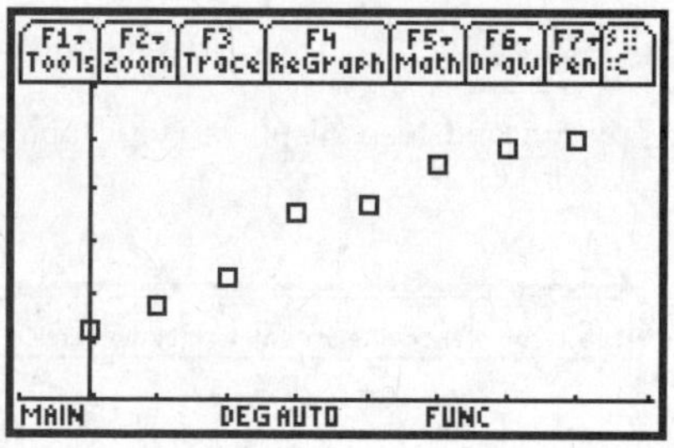

To turn off the plot, highlight Plot 1 on the equation-editor screen and then press $\boxed{\text{F4}}$. Note that there is no check mark to the left of Plot 1 when it is turned off.

Now we will fit a linear equation to the data. From the data list screen, press $\boxed{\text{F5}}$ to display the Calculate menu. Press $\boxed{\triangleright}$ 5 to select LinReg (linear regression). Then press $\boxed{\triangledown}$ $\boxed{\text{alpha}}$ $\boxed{\text{C}}$ 1 $\boxed{\triangledown}$ $\boxed{\text{alpha}}$ $\boxed{\text{C}}$ 2 to indicate that the data in c1 and c2 will be used for x and y, respectively. Press $\boxed{\triangledown}$ $\boxed{\triangleright}$ $\boxed{\triangledown}$ $\boxed{\text{ENTER}}$ to indicate that the regression equation should be copied to the equation-editor screen as y_1. Finally press $\boxed{\text{ENTER}}$ again to see the STAT VARS screen which displays the coefficients a and b of the regression equation $y = ax + b$.

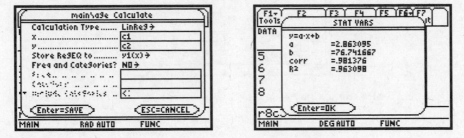

Note that values for "corr" (the correlation coefficient) and r^2 (the coefficient of determination) will also be displayed. These numbers indicate how well the regression line fits the data. While it is possible to suppress these numbers on some graphing calculators, this cannot be done on the TI-89.

To estimate the number of U. S. households subscribing to cable television in 2010, evaluate the regression equation for $x = 11$. (2010 is 11 years after 1999.) Use any of the methods for evaluating a function presented earlier in this chapter. (See pages 89 - 91 of this manual.) We will use function notation on the home screen. When $x = 11, y \approx 108.2$, so we predict that there will be about 108.2 million U. S. households subscribing to cable television in 2010.

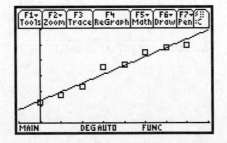

We can also plot the data points along with the graph of the regression equation. To do this we first define a plot as described on pages 92 and 93 of this manual. Now select a viewing window. Press $\boxed{\diamond}$ $\boxed{\text{WINDOW}}$ to go to the WINDOW screen. Enter the desired dimensions or press $\boxed{\text{F2}}$ 9 to activate the ZoomData operation which automatically defines a viewing window that displays all of the points and also displays the graph. Here we continue to use the dimensions $[-1, 8, 70, 100]$, xscl $= 1$, yscl $= 5$.

Turn off the plot as described on page 93 of this manual before graphing other functions.

THE INTERSECT METHOD

We can use the Intersection feature from the Math menu on the GRAPH screen to solve equations. We call this the **Intersect method**.

Section 1.5, Example 1 Solve $\frac{3}{4}x - 1 = \frac{7}{5}$.

Press $\boxed{\diamond}$ $\boxed{Y=}$ to go to the equation-editor screen. Clear any existing entries and then enter $y_1 = \frac{3}{4}x - 1$ and $y_2 = \frac{7}{5}$. The solution of the original equation is the first coordinate of the point of intersection of the graphs of y_1 and y_2. Graph the equations in a window that displays the point of intersection. The window $[-5, 5, -5, 5]$ is a good choice.

We will use the Intersection feature to find the coordinates of the point of intersection. Press $\boxed{F5}$ 5 to select Intersection from the Math menu on the Graph screen. The query "1st curve?" appears at the bottom of the screen. The blinking cursor is positioned on the graph of y_1. This is indicated by the 1 in the upper right-hand corner of the screen. Press $\boxed{\text{ENTER}}$ to indicate that this is the first curve involved in the intersection. Next the query "2nd curve?" appears at the bottom of the screen. The blinking cursor is now positioned on the graph of y_2 and the notation 2 should appear in the top right-hand corner of the screen. Press $\boxed{\text{ENTER}}$ to indicate that this is the second curve. We identify the curves for the calculator since we could have more than two graphs on the screen at once. After we identify the second curve, the query "Lower bound?" appears at the bottom of the screen. Use the left or right arrow key to move the blinking cursor to a point to the left of the point of intersection of the lines or type an x-value less than the x-coordinate of the point of intersection. Then press $\boxed{\text{ENTER}}$. Next the query "Upper bound?" appears. Move the cursor to a point to the right of the point of intersection or type an x-value greater than the x-value of the point of intersection and press $\boxed{\text{ENTER}}$. We give a lower and an upper bound since some pairs of curves have more than one point of intersection. Now the coordinates of the point of intersection appear at the bottom of the screen. We see that the first coordinate of the point of intersection is 3.2, so the solution of the equation is 3.2, or $\frac{16}{5}$.

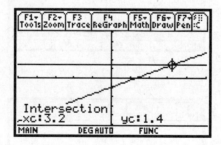

THE ZERO METHOD

When an equation is expressed in the form $f(x) = 0$, it can be solved using the Zero feature from the Math menu on the Graph screen.

Section 1.5, Example 11 Find the zero of $f(x) = 5x - 9$.

On the equation-editor screen, clear any existing entries and then enter $y_1 = 5x - 9$. Now graph the function in a viewing window that shows the x-intercept clearly. The standard window is a good choice.

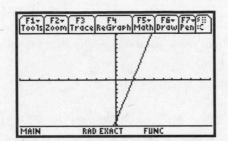

Press [F5] 2 to select the Zero feature from the Math menu. We are prompted to select a lower bound. This means that we must choose an x-value that is to the left of the x-intercept. This can be done by using the left- and right-arrow keys to move to a point on the curve to the left of the x-intercept or by keying in a value less than the x-coordinate of the intercept.

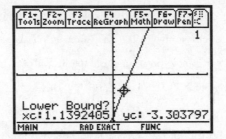

Once this is done press [ENTER]. Now we are prompted to select an upper bound that is to the right of the x-intercept. Again, this can be done by using the arrow keys to move to a point on the curve to the right of the x-intercept or by keying in a value greater than the x-coordinate of the intercept.

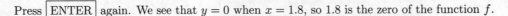

Press [ENTER] again. We see that $y = 0$ when $x = 1.8$, so 1.8 is the zero of the function f.

Chapter 2
More on Functions

THE MAXIMUM AND MINIMUM FEATURES

Section 2.1, Example 2 Use a graphing calculator to determine any relative maxima or minima of the function $f(x) = 0.1x^3 - 0.6x^2 - 0.1x + 2$.

First graph $y_1 = 0.1x^3 - 0.6x^2 - 0.1x + 2$ in a window that displays the relative extrema of the function. Trial and error reveals that one good choice is $[-4, 6, -3, 3]$. Observe that a relative maximum occurs near $x = 0$ and a relative minimum occurs near $x = 4$.

To find the relative maximum, first press $\boxed{\text{F5}}$ $\boxed{4}$ or $\boxed{\text{F5}}$ $\boxed{\triangledown}$ $\boxed{\triangledown}$ $\boxed{\triangledown}$ $\boxed{\text{ENTER}}$ to select the Maximum feature from the Math menu on the Graph screen. We are prompted to select a lower bound for the relative maximum. This means that we must choose an x-value that is to the left of the x-value of the point where the relative maximum occurs. This can be done by using the left- and right-arrow keys to move the cursor to a point to the left of the relative maximum or by keying in an appropriate value.

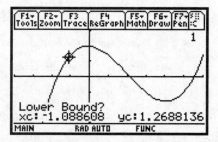

Once this is done, press $\boxed{\text{ENTER}}$. Now we are prompted to select an upper bound. We move the cursor to a point to the right of the relative maximum or we key in an appropriate value.

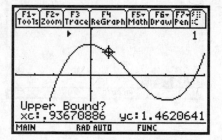

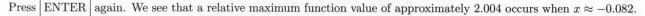

Press $\boxed{\text{ENTER}}$ again. We see that a relative maximum function value of approximately 2.004 occurs when $x \approx -0.082$.

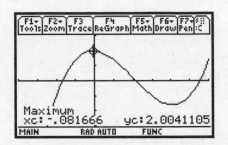

To find the relative minimum, select the Minimum feature from the Math menu by pressing $\boxed{\text{F5}}$ 3 or $\boxed{\text{F5}}$ $\boxed{\triangledown}$ $\boxed{\triangledown}$ $\boxed{\text{ENTER}}$. Select lower and upper bounds for the relative minimum as described above. We see that a relative minimum function value of approximately -1.604 occurs when $x \approx 4.082$.

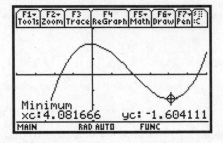

THE GREATEST INTEGER FUNCTION

The greatest integer function is found in the Catalog and is denoted "int." To find int(1.9) first press $\boxed{\text{CATALOG}}$ $\boxed{\text{I}}$ to go to the first item in the Catalog that begins with I. (I is the purple alphabetic operation associated with the 9 numeric key.) Note that it is not necessary to press $\boxed{\text{alpha}}$ before $\boxed{\text{I}}$ when the Catalog is displayed. Now move the triangular selection cursor beside "int(" and press $\boxed{\text{ENTER}}$ to copy "int(" to the entry line of the home screen. Then press 1 $\boxed{.}$ 9 $\boxed{)}$ $\boxed{\text{ENTER}}$. We can also type "int" directly on the entry line of the home screen. Press $\boxed{\text{2nd}}$ $\boxed{\text{a-lock}}$ $\boxed{\text{I}}$ $\boxed{\text{N}}$ $\boxed{\text{T}}$ $\boxed{\text{alpha}}$ to do this. Then finish entering int(1.9) by pressing $\boxed{(}$ 1 $\boxed{.}$ 9 $\boxed{)}$ $\boxed{\text{ENTER}}$.

We can also graph the greatest integer function.

Section 2.1, Example 9 Graph $f(x) = \text{int}(x)$.

First enter the equation on the equation-editor screen. To do this, clear any existing entries as described on page 85 of this manual and then position the cursor beside y1 =. Copy "int(" from the catalog as described above. Then press $\boxed{\text{X}}$ $\boxed{)}$ $\boxed{\text{ENTER}}$.

Now we set the calculator in the Dot graph style. In the Line graph style, which we have used previously, the calculator will connect adjacent points, or pixels, with line segments and produce a graph that looks like stair-steps rather than a set of horizontal line segments. To select the Dot graph style, highlight the equation and press $\boxed{\text{2nd}}$ $\boxed{\text{F6}}$ 2 or $\boxed{\text{2nd}}$ $\boxed{\text{F6}}$ $\boxed{\triangledown}$ $\boxed{\text{ENTER}}$.

Now choose a window and press $\boxed{\diamond}$ $\boxed{\text{GRAPH}}$. The window $[-6, 6, -6, 6]$ is shown here.

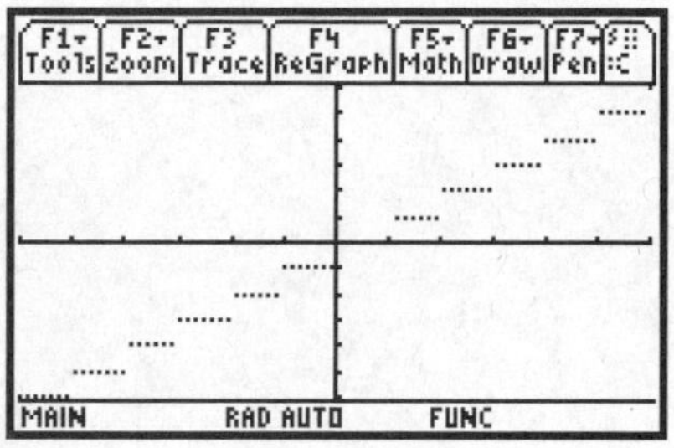

THE COMPOSITION OF FUNCTIONS

We can evaluate composite functions on a graphing calculator.

Section 2.3, Example 1 (b) Given that $f(x) = 2x - 5$ and $g(x) = x^2 - 3x + 8$, find $(f \circ g)(7)$ and $(g \circ f)(7)$.

On the equation-editor screen enter $y_1 = 2x - 5$ and $y_2 = x^2 - 3x + 8$. Then $(f \circ g)(7) = (y_1 \circ y_2)(7)$, or $y_1(y_2(7))$ and $(g \circ f)(7) = (y_2 \circ y_1)(7)$, or $y_2(y_1(7))$. To find these function values press $\boxed{\text{HOME}}$ or $\boxed{\text{2nd}}$ $\boxed{\text{QUIT}}$ to go to the home screen. Then enter $y_1(y_2(7))$ by pressing $\boxed{\text{Y}}$ 1 $\boxed{(}$ $\boxed{\text{Y}}$ 2 $\boxed{(}$ 7 $\boxed{)}$ $\boxed{)}$ $\boxed{\text{ENTER}}$. Enter $y_2(y_1(7))$ by pressing $\boxed{\text{Y}}$ 2 $\boxed{(}$ $\boxed{\text{Y}}$ 1 $\boxed{(}$ 7 $\boxed{)}$ $\boxed{)}$ $\boxed{\text{ENTER}}$.

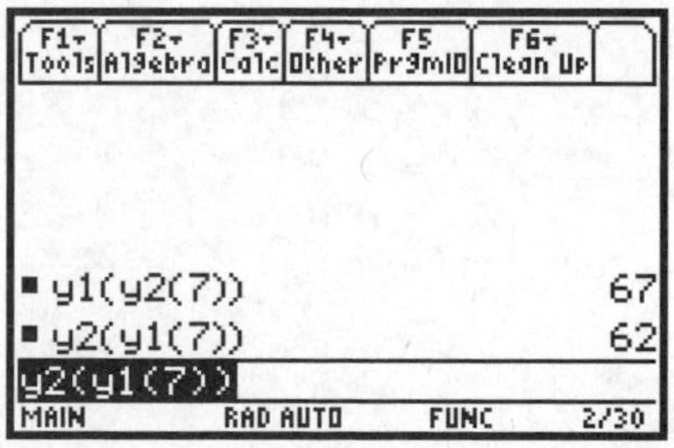

Chapter 3
Quadratic Functions and Equations; Inequalities

OPERATIONS WITH COMPLEX NUMBERS

Operations with complex numbers can be performed on the TI-89. First set the calculator in the rectangular mode by pressing ⎢MODE⎥, highlighting the entry for Complex Format, and then pressing ⎢▷⎥ 2 ⎢ENTER⎥ or ⎢▷⎥ ⎢▽⎥ ⎢ENTER⎥ ⎢ENTER⎥.

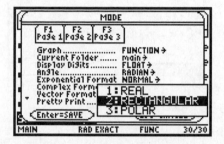

Section 3.1, Example 2

(a) Add: $(8 + 6i) + (3 + 2i)$.

To find this sum go to the home screen and press 8 ⎢+⎥ 6 ⎢2nd⎥ ⎢i⎥ ⎢+⎥ 3 ⎢+⎥ 2 ⎢2nd⎥ ⎢i⎥ ⎢ENTER⎥. (The number i is the second operation associated with the ⎢CATALOG⎥ key.) Note that it is not necessary to include parentheses when we are adding.

(b) Subtract: $(4 + 5i) - (6 - 3i)$.

Press 4 ⎢+⎥ 5 ⎢2nd⎥ ⎢i⎥ ⎢-⎥ ⎢(⎥ 6 ⎢-⎥ 3 ⎢2nd⎥ ⎢i⎥ ⎢)⎥ ⎢ENTER⎥. Note that the parentheses must be included as shown so that the entire number $6 - 3i$ is subtracted.

Section 3.1, Example 3

(a) Multiply: $\sqrt{-16} \cdot \sqrt{-25}$.

   Press $\boxed{\text{2nd}}$ $\boxed{\sqrt{}}$ $\boxed{(-)}$ 1 6 $\boxed{)}$ $\boxed{\text{2nd}}$ $\boxed{\sqrt{}}$ $\boxed{(-)}$ 2 5 $\boxed{)}$ $\boxed{\text{ENTER}}$.

(b) Multiply: $(1 + 2i)(1 + 3i)$.

   Press $\boxed{(}$ 1 $\boxed{+}$ 2 $\boxed{\text{2nd}}$ $\boxed{i}$ $\boxed{)}$ $\boxed{(}$ 1 $\boxed{+}$ 3 $\boxed{\text{2nd}}$ $\boxed{i}$ $\boxed{)}$ $\boxed{\text{ENTER}}$.

(c) Multiply: $(3 - 7i)^2$.

   Press $\boxed{(}$ 3 $\boxed{-}$ 7 $\boxed{\text{2nd}}$ $\boxed{i}$ $\boxed{)}$ $\boxed{\wedge}$ 2 $\boxed{\text{ENTER}}$.

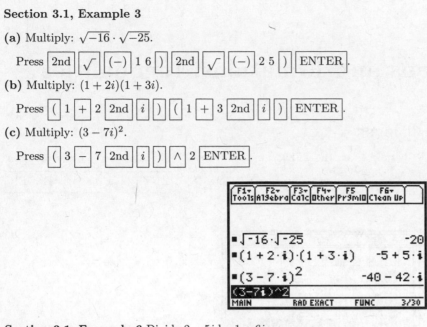

Section 3.1, Example 6 Divide $2 - 5i$ by $1 - 6i$.

   We have $\dfrac{2 - 5i}{1 - 6i}$, or $(2 - 5i) \div (1 - 6i)$. Press $\boxed{(}$ 2 $\boxed{-}$ 5 $\boxed{\text{2nd}}$ $\boxed{i}$ $\boxed{)}$ $\boxed{\div}$ $\boxed{(}$ 1 $\boxed{-}$ 6 $\boxed{\text{2nd}}$ $\boxed{i}$ $\boxed{)}$ $\boxed{\text{ENTER}}$.

Chapter 4
Polynomial and Rational Functions

POLYNOMIAL MODELS

We can fit third-degree, or cubic, functions and fourth-degree, or quartic, functions to data on a TI-89.

Section 4.1, Example 9 The table below shows the number of miles of U. S.-owned operating railroad track for several years beginning with 1830. Model the data with a cubic function and with a quartic function. Let the first coordinate of each data point be the number of years after 1830.

Years, x	Miles of U. S.-Owned Operating Railroad Track
1830, 0	23
1840, 10	2,818
1850, 20	9,021
1860, 30	30,635
1870, 40	52,922
1880, 50	92,147
1890, 60	163,597
1900, 70	193,346
1910, 80	240,293
1916, 86	254,037
1920, 90	252,845
1930, 100	249,052
1940, 110	233,670
1950, 120	223,779
1960, 130	217,552
1970, 140	205,782
1980, 150	178,056
1990, 160	145,979
2000, 170	144,473
2003, 173	141,509
2005, 175	140,810

Enter the data in the Data/Matrix editor as described on page 92 of this manual. To model the data with a cubic function, select cubic regression from the Calculate menu after the data are entered. Do this from the data-entry screen by pressing F5 to display the calculate screen. Then press $\boxed{\triangleright}$ 3 $\boxed{\text{ENTER}}$. Indicate that the data in c1 and c2 will be used for x and y, respectively, and that the equation should be copied to the equation-editor screen as y_1 as described on pages 92 and 93 of this manual. The calculator displays the coefficients of a cubic function $y = ax^3 + bx^2 + cx + d$.

To model the data with a quartic function, select quartic regression from the Calculate menu after the data are entered. Press $\boxed{\text{F5}}$ $\boxed{\triangleright}$ $\boxed{\text{alpha}}$ $\boxed{\text{A}}$ $\boxed{\text{ENTER}}$. Indicate that the data in c1 and c2 will be used for x and y, respectively, and that the equation should be copied to the equation-editor screen as y_1 as described on pages 92 and 93 of this manual. The calculator displays the coefficients of a quartic function $y = ax^4 + bx^3 + cx^2 + dx + e$.

A scatterplot of the data can be graphed as described on page 92 of this manual. The function has been copied to the Y = screen, so it can be graphed along with the scatterplot. It can also be evaluated using one of the methods on pages 89 - 91.

Chapter 5
Exponential and Logarithmic Functions

GRAPHING AN INVERSE FUNCTION

The DrawInv operation can be used to graph a function and its inverse on the same screen. A formula for the inverse function need not be found in order to do this. The calculator must be set in Func mode when this operation is used.

Section 5.1, Example 7 Graph $f(x) = 2x - 3$ and $f^{-1}(x)$ using the same set of axes.

Enter $y_1 = 2x - 3$, clear all other functions on the Y = screen, and choose a viewing window. Press $\boxed{\text{HOME}}$ to go to the home screen. Then press $\boxed{\text{CATALOG}}$ $\boxed{\text{D}}$ to go to the first item in the Catalog that begins with D. Use the $\boxed{\triangledown}$ key to scroll down to DrawInv and then press $\boxed{\text{ENTER}}$ to copy DrawInv to the entry line of the home screen. Follow these keystrokes with $\boxed{\text{Y}}$ 1 $\boxed{(}$ $\boxed{\text{X}}$ $\boxed{)}$ to select function y_1. Press $\boxed{\text{ENTER}}$ to see the graph of the function and its inverse. The graphs are shown here in the standard window.

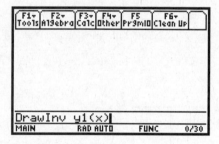 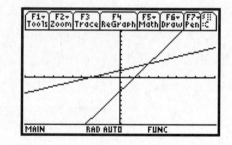

EVALUATING e^x, Log x, and Ln x

We can use the calculator's scientific keys to evaluate e^x, $\log x$, and $\ln x$ for specific values of x.

Section 5.2, Example 5 (a), (b) Find the values of e^3 and $e^{-0.23}$. Round to four decimal places.

First select Approximate mode by pressing $\boxed{\text{MODE}}$ $\boxed{\text{F2}}$ $\boxed{\triangledown}$ $\boxed{\triangledown}$ $\boxed{\triangleright}$ 3 $\boxed{\text{ENTER}}$. Then to find e^3 press $\boxed{\diamond}$ $\boxed{e^x}$ 3 $\boxed{)}$ $\boxed{\text{ENTER}}$. (e^x is the green $\diamond$ operation associated with the $\boxed{\text{X}}$ key.) The calculator returns 20.0855369232. Thus, $e^3 \approx$ 20.0855. To find $e^{-0.23}$ press $\boxed{\diamond}$ $\boxed{e^x}$ $\boxed{(-)}$ $\boxed{\cdot}$ $2\,3$ $\boxed{)}$ $\boxed{\text{ENTER}}$. The calculator returns .794533602503, so $e^{-0.23} \approx 0.7945$.

Section 5.3, Example 5 Find the values of log 645,778, log 0.0000239, and log (−3). Round to four decimal places.

The calculator must be in Approximate mode as described in Example 6 above. To find log 645,778 press $\boxed{\text{2nd}}$ $\boxed{\text{a-lock}}$ $\boxed{\text{L}}$ $\boxed{\text{O}}$ $\boxed{\text{G}}$ $\boxed{\text{alpha}}$ $\boxed{(}$ $6\,4\,5\,7\,7\,8$ $\boxed{)}$ $\boxed{\text{ENTER}}$ and read 5.81008324563. Thus, $\log 645,778 \approx 5.8101$. The operation "log(" can also be selected from the Catalog. To find log 0.0000239 press $\boxed{\text{2nd}}$ $\boxed{\text{a-lock}}$ $\boxed{\text{L}}$ $\boxed{\text{O}}$ $\boxed{\text{G}}$ $\boxed{\text{alpha}}$ $\boxed{(}$ $\boxed{\cdot}$ $0\,0\,0\,0\,2\,3\,9$ $\boxed{)}$ $\boxed{\text{ENTER}}$. The calculator returns −4.62160209905, so $\log 0.0000239 \approx -4.6216$. The previous entry, log 645,778, can also be edited to find log 0.0000239. When the TI-89 has Real selected for the Complex Format mode the keystrokes $\boxed{\text{2nd}}$

$\boxed{\text{a-lock}}$ $\boxed{\text{L}}$ $\boxed{\text{O}}$ $\boxed{\text{G}}$ $\boxed{\text{alpha}}$ $\boxed{(}$ $\boxed{(-)}$ 3 $\boxed{)}$ $\boxed{\text{ENTER}}$ produce the error message "Non-real result," indicating that the result of this calculation is not a real number. We could also have entered log(−3) by editing the previous entry.

Section 5.3, Example 6 (a), (b), (c) Find the values of ln 645,778, ln 0.0000239, and ln (−5). Round to four decimal places.

The calculator must be in Approximate mode as described in Example 6 above. To find ln 645,778 and ln 0.0000239 repeat the keystrokes used above to find log 645,778 and log 0.0000239 but press $\boxed{\text{2nd}}$ $\boxed{\text{LN}}$ rather than $\boxed{\text{2nd}}$ $\boxed{\text{a-lock}}$ $\boxed{\text{L}}$ $\boxed{\text{O}}$ $\boxed{\text{G}}$ $\boxed{\text{alpha}}$ $\boxed{(}$. (LN is the second operation associated with the $\boxed{\text{X}}$ key.) We find that ln 645,778 ≈ 13.3782 and ln 0.0000239 ≈ −10.6416. When the TI-89 has Real selected for the Complex Format mode the keystrokes $\boxed{\text{2nd}}$ $\boxed{\text{LN}}$ $\boxed{(-)}$ 5 $\boxed{)}$ $\boxed{\text{ENTER}}$ produce the error message "Non-real result," indicating that the result of this calculation is not a real number.

USING THE CHANGE OF BASE FORMULA

To find a logarithm with a base other than 10 or e we use the change-of-base formula, $\log_b M = \dfrac{\log_a M}{\log_a b}$, where a and b are any logarithmic bases and M is any positive number.

Section 5.3, Example 7 Find $\log_5 8$ using common logarithms.

The calculator must be in Approximate mode as described in Example 6 above. We let $a = 10$, $b = 5$, and $M = 8$ and substitute in the change-of-base formula. We have $\log_5 8 = \dfrac{\log_{10} 8}{\log_{10} 5}$. To carry out the division, press $\boxed{\text{2nd}}$ $\boxed{\text{a-lock}}$ $\boxed{\text{L}}$ $\boxed{\text{O}}$ $\boxed{\text{G}}$ $\boxed{\text{alpha}}$ $\boxed{(}$ 8 $\boxed{)}$ $\boxed{\div}$ $\boxed{\text{2nd}}$ $\boxed{\text{a-lock}}$ $\boxed{\text{L}}$ $\boxed{\text{O}}$ $\boxed{\text{G}}$ $\boxed{\text{alpha}}$ $\boxed{(}$ 5 $\boxed{)}$ $\boxed{\text{ENTER}}$. The result is about 1.2920. We could have let $a = e$ and used natural logarithms to find $\log_5 8$ as well.

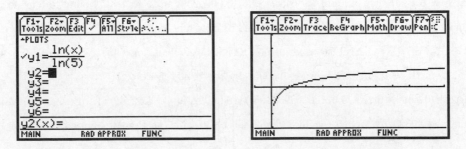

Section 5.3, Example 9 Graph $y = \log_5 x$.

To use a TI-89 we must first change the base to e or 10. Here we use e. Let $a = e$, $b = 5$, and $M = x$ and substitute in the change-of-base formula. Enter $y_1 = \dfrac{\ln x}{\ln 5}$ on the Y = screen, select a window, and press $\boxed{\diamond}$ $\boxed{\text{GRAPH}}$.

EXPONENTIAL AND LOGARITHMIC REGRESSION

In addition to the types of polynomial regression discussed earlier, exponential and logarithmic functions can be fit to data. The operations of entering data, making scatterplots, and graphing and evaluating these functions are the same as for linear regression functions. So are the procedures for copying a regression equation to the Y = screen, graphing it, and using it to find function values.

Section 5.6, Example 6 *Surveillance Cameras.* The number of U. S. communities using surveillance cameras at intersections has increased greatly in recent years, as shown in the following table.

Year, x	Number of U. S. Communities Using Surveillance Cameras
1999, 0	19
2001, 2	35
2003, 4	75
2005, 6	130
2007, 8	243

(a) Use a graphing calculator to fit an exponential function to the data.

Enter the data in the Data/Matrix editor as described on page 92 of this manual. Then select exponential regression from the Calculate menu by pressing F5 ▷ 4 ENTER. Indicate that the data in c1 and c2 will be used for x and y, respectively, and that the equation should be copied to the equation-editor screen as y_1 as described on pages 92 and 93 of this manual. The calculator displays the coefficient a and the base b for the exponential function $y = a \cdot b^x$.

A scatterplot of the data can be graphed as described on pages 92 and 93 of this manual. The function has been copied to the Y = screen, so it can be graphed along with the scatterplot. It can also be evaluated using one of the methods on pages 89 - 91.

Section 5.6, Exercise 30 *Forgetting.* In an art class, students were tested at the end of the course on a final exam. Then they were retested with an equivalent test at subsequent time intervals. Their scores after time x, in months, are given in the following table. Use a graphing calculator to fit a logarithmic function $y = a + b\ln(x)$ to the data.

Time, x (in months)	Score, y
1	84.9%
2	84.6%
3	84.4%
4	84.2%
5	84.1%
6	83.9%

After entering the data in the Data/Matrix editor as described on page 92 of this manual, press F5 ▷ 6 ENTER to go to the Calculate menu and select LnReg. Indicate that the data in c1 and c2 will be used for x and y, respectively, and that the equation should be copied to the equation-editor screen as y_1 as described on pages 92 and 93 of this manual. The values of a and b for the logarithmic function $y = a + b\ln(x)$ are displayed. The function can be evaluated using one of the methods on pages 89 - 91.

LOGISTIC REGRESSION

A logistic function can be fit to data using the TI-89.

Section 5.6, Exercise 34 *Effect of Advertising.* A company introduces a new software product on a trial run in a city. They advertised the product on television and found the following data relating the percent P of people who bought the product after x ads were run. Use a graphing calculator to fit a logistic function to the data.

Number of Ads, x	Percent Who Bought, P
0	0.2
10	0.7
20	2.7
30	9.2
40	27
50	57.6
60	83.3
70	94.8
80	98.5
90	99.6

After entering the data in the Data/Matrix editor as described on page 92 of this manual, press F5 ▷ alpha C to go to the Calculate menu and select Logistic. Indicate that the data in c1 and c2 will be used for x and y, respectively, and that the equation should be copied to the equation-editor screen as y_1 as described on pages 92 and 93 of this manual. The values of a, b, and c for the logistic function $y = a/(1 + be^{cx}) + d$ are displayed.

```
┌─────────────────────────────────────────┐
│ ╱F1·╲ ╱ F2 ╲╱ F3 ╲╱ F4 ╲╱ F5 ╲╱F6·╲╱F7 │
│ │Tools│        STAT VARS            │ lt │
│ │DATA │                                  │
│ │     │ y=a/(1+b·e^(c·x))+d              │
│ │     │  a      =99.995142              │
│ │ 8   │  b      =488.75131              │
│ │ 9   │  c      =-.129971               │
│ │ 10  │  d      =-.005074               │
│ │ 11  │                                  │
│ │     │                                  │
│ │r11  │ ⟨ Enter=OK ⟩                    │
│ │MAIN │     RAD APPROX    FUNC          │
└─────────────────────────────────────────┘
```

Chapter 6
The Trigonometric Functions

CONVERTING BETWEEN D°M′S″ AND DECIMAL DEGREE MEASURE

We can convert decimal notation to D°M′S″ notation and vice versa on the TI-89.

Section 6.1, Example 5 Convert 5°42′30″ to decimal degree notation.

Select Degree for the Angle mode setting and Approximate for the Exact/Approx setting, Then press 5 ⟨2nd⟩ ⟨°⟩ 4 2 ⟨2nd⟩ ⟨′⟩ 3 0 ⟨2nd⟩ ⟨″⟩ ⟨ENTER⟩. (°, ′, and ″ are the second operations associated with the ⟨ | ⟩, ⟨ = ⟩, and 1 keys, respectively.) The calculator returns 5.70833333333, so 5°42′30″ ≈ 5.71°.

```
/ F1▾ \ F2▾ \ F3▾ \ F4▾ \ F5 \ F6▾ \
 Tools Algebra Calc Other Prgmi0 Clean Up


 ■5°42'30"      5.70833333333
 5°42'30"
 MAIN        DEG APPROX    POL    1/30
```

Section 6.1, Exercise 39 Convert 15′5″ to decimal degree notation.

In converting from D°M′S″ notation to decimal degree notation, we must always enter a number for degrees even if it is 0. Thus, to convert 15′5″ to decimal degree notation, we enter 0°15′5″ as described above. The result is approximately 0.25°.

Section 6.1, Exercise 41 Convert 5°53″ to decimal degree notation.

All entries containing a nonzero number of seconds must include entries for the number of degrees and minutes even if one or both is 0. Thus, to convert 5°53″ to decimal degree notation, we enter 5°0′53″ as described in Example 5 above. The result is approximately 5.01°.

```
/ F1▾ \ F2▾ \ F3▾ \ F4▾ \ F5 \ F6▾ \
 Tools Algebra Calc Other Prgmi0 Clean Up


 ■0°15'5"        .251388888889
 ■5°0'53"       5.01472222222
 5°0'53"
 MAIN        DEG APPROX    FUNC
```

Section 6.1, Example 6 Convert 72.18°to D°M′S″ notation.

Select Degree for the Angle mode setting. Then press 7 2 $\boxed{\cdot}$ 1 8 $\boxed{\text{2nd}}$ $\boxed{\text{MATH}}$ $\boxed{\triangledown}$ $\boxed{\triangleright}$ to enter the angle and display the MATH Angle menu. (MATH is the second operation associated with the 5 numeric key.) Press 8 to select ▷DMS and then press $\boxed{\text{ENTER}}$ to see D°M′S″ notation for the angle. The calculator returns 72°10′48″.

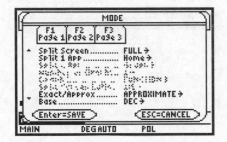

FINDING TRIGONOMETRIC FUNCTION VALUES

A graphing calculator's SIN, COS, and TAN operations can be used to find the values of trigonometric functions. When angles are given in degree measure, Degree must be selected for the Angle mode as in the example above. We must also select Approximate for the Exact/Approx mode setting.

Section 6.1, Example 7 Find the trigonometric function value, rounded to four decimal places, of each of the following.

a) tan 29.7°   b) sec 48°   c) sin 84°10′39″

a) Press $\boxed{\text{2nd}}$ $\boxed{\text{TAN}}$ 2 9 $\boxed{\cdot}$ 7 $\boxed{)}$ $\boxed{\text{ENTER}}$. (TAN is the second operation associated with the $\boxed{\text{T}}$ key.) We find that tan 29.7° ≈ 0.5704.

b) The secant, cosecant, and cotangent functions can be found by taking the reciprocals of the cosine, sine, and tangent functions, respectively. To find sec 48° we enter the reciprocal of cos 48° by pressing 1 $\boxed{\div}$ $\boxed{\text{2nd}}$ $\boxed{\text{COS}}$ 4 8 $\boxed{)}$ $\boxed{\text{ENTER}}$. (COS is the second operation associated with the $\boxed{\text{Z}}$ key.) We can also find sec 48° by entering $(\cos(48))^{-1}$. To do this press $\boxed{(}$ $\boxed{\text{2nd}}$ $\boxed{\text{COS}}$ 4 8 $\boxed{)}$ $\boxed{)}$. Then select $\wedge - 1$ from the Catalog or press $\boxed{\wedge}$ $\boxed{(-)}$ 1. Finally press $\boxed{\text{ENTER}}$. The result is sec 48° ≈ 1.4945.

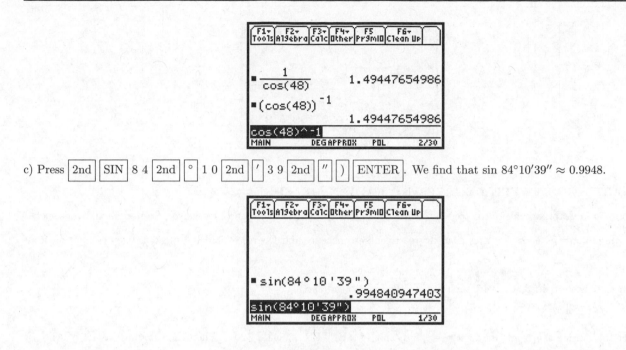

c) Press $\boxed{\text{2nd}}$ $\boxed{\text{SIN}}$ 8 4 $\boxed{\text{2nd}}$ $\boxed{\circ}$ 1 0 $\boxed{\text{2nd}}$ $\boxed{'}$ 3 9 $\boxed{\text{2nd}}$ $\boxed{''}$ $\boxed{)}$ $\boxed{\text{ENTER}}$. We find that $\sin 84°10'39'' \approx 0.9948$.

FINDING ANGLES

The inverse trigonometric function keys provide a quick way to find an angle given a trigonometric function value for that angle.

Section 6.1, Example 8 Find the acute angle, to the nearest tenth of a degree, whose sine value is approximately 0.20113.

Although the TABLE feature can be used to approximate this angle, it is faster to use the inverse sine key. With the calculator set in Degree and Approximate modes, press $\boxed{\diamond}$ $\boxed{\text{SIN}^{-1}}$ $\boxed{\cdot}$ 2 0 1 1 3 $\boxed{)}$ $\boxed{\text{ENTER}}$. (SIN^{-1} is the green $\diamond$ operation associated with the $\boxed{\text{Y}}$ key.) We find that the desired acute angle is approximately $11.6°$.

Section 6.1, Exercise 63 Find the acute angle, to the nearest tenth of a degree, whose cotangent value is 2.127.

Angles whose secant, cosecant, or cotangent values are known can be found using the reciprocals of the cosine, sine, and tangent functions, respectively. Since $\cot\theta = \dfrac{1}{\tan\theta} = 2.127$, we have $\tan\theta = \dfrac{1}{2.127}$, or $(2.127)^{-1}$. To find θ press $\boxed{\diamond}$ $\boxed{\text{TAN}^{-1}}$ 1 $\boxed{\div}$ 2 $\boxed{\cdot}$ 1 2 7 $\boxed{)}$ $\boxed{\text{ENTER}}$ or $\boxed{\diamond}$ $\boxed{\text{TAN}^{-1}}$ 2 $\boxed{\cdot}$ 1 2 7 $\wedge$−1 $\boxed{)}$ $\boxed{\text{ENTER}}$. (TAN^{-1} is the green $\diamond$ operation associated with the $\boxed{\text{T}}$ key. $\wedge$−1 is entered by pressing $\boxed{\wedge}$ $\boxed{(-)}$ 1 or can be selected from the Catalog.) We find that $\theta \approx 25.2°$.

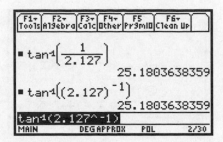

CONVERTING BETWEEN DEGREE AND RADIAN MEASURE

We can use a graphing calculator to convert from degree to radian measure and vice versa. Radian should be selected for the Angle mode when converting from degree to radian measure and Degree should be selected when converting from radian to degree measure.

Section 6.4, Example 3 Convert each of the following to radians.

a) 120°   b) −297.25°

a) Select Radian for the Angle mode. Press 1 2 0 $\boxed{\text{2nd}}$ $\boxed{\circ}$ $\boxed{\text{ENTER}}$ to enter 120°. When the calculator is set in Auto or Exact mode, it returns $2\pi/3$ radians.

b) With Radian selected for the Angle mode press $\boxed{(-)}$ 2 9 7 $\boxed{\cdot}$ 2 5 $\boxed{\text{2nd}}$ $\boxed{\circ}$ $\boxed{\text{ENTER}}$. We see that −297.25° ≈ −5.19 radians.

Section 6.4, Example 4 Convert each of the following to degrees.

a) $\dfrac{3\pi}{4}$ radians   b) 8.5 radians

a) Select Degree for the Angle mode. Then press $\boxed{(}$ 3 $\boxed{\text{2nd}}$ $\boxed{\pi}$ $\boxed{\div}$ 4 $\boxed{)}$ $\boxed{\text{2nd}}$ $\boxed{\text{MATH}}$ $\boxed{\triangledown}$ $\boxed{\triangleright}$ 2 $\boxed{\text{ENTER}}$ to enter $\dfrac{3\pi}{4}$ radians. (π is the second operation associated with the $\boxed{\wedge}$ key). The calculator returns 135, so $3\pi/4$ radians = 135°. Note that the parentheses are necessary in order to enter the entire expression in radian measure. Without the parentheses, the calculator reads only the denominator, 4, in radian measure and an incorrect result occurs.

b) With the calculator set in Degree mode press 8 $\boxed{\cdot}$ 5 $\boxed{\text{2nd}}$ $\boxed{\text{MATH}}$ $\boxed{\triangledown}$ $\boxed{\triangleright}$ 2 $\boxed{\text{ENTER}}$. The calculator returns 487.0141259, so 8.5 radians ≈ 487.01°.

```
┌─────────────────────────────────┐
│             MODE                │
│ ┌────┬────┬────┐                │
│ │ F1 │ F2 │ F3 │                │
│ │Page 1│Page 2│Page 3│          │
│                                 │
│ Graph............... FUNCTION→   │
│ Current Folder...... main→       │
│ Display Digits...... FLOAT→      │
│ Angle............... DEGREE→     │
│ Exponential Format   NORMAL→     │
│ Complex Format..... RECTANGULAR→ │
│ Vector Format...... RECTANGULAR→ │
│ Pretty Print....... ON→          │
│                                 │
│  ⟨Enter=SAVE⟩      ⟨ESC=CANCEL⟩  │
│ MAIN      RAD APPROX   FUNC      │
└─────────────────────────────────┘
```

```
┌─────────────────────────────────┐
│┌──┬───┬──┬───┬──┬────────┐        │
││F1▾│F2▾│F3▾│F4▾│F5│F6▾    │       │
││Tools│Algebra│Calc│Other│PrgmIO│Clean Up│ │
│                                 │
│  ■ ⎛ 3·π ⎞ʳ                      │
│    ⎜─────⎟              135.      │
│    ⎝  4  ⎠                        │
│                                 │
│  ■ (8.5)ʳ        487.014125861   │
│ ┌───────────────────────────────┐│
│ │8.5ʳ                           ││
│ └───────────────────────────────┘│
│ MAIN    DEG APPROX   FUNC   2/30  │
└─────────────────────────────────┘
```

Chapter 7
Trigonometric Identities, Inverse Functions, and Equations

THE PATH GRAPH STYLE

Graph styles can be selected from the Style menu on the equation-editor screen of the TI-89. The path graph style can be used, along with the line style, to determine whether graphs coincide. This can be used to provide a partial check of an identity.

Section 7.1, Example 1 Use a graphing calculator to do a partial check of the identity $\cos x(\tan x - \sec x) = \sin x - 1$.

First, on the Y = screen, enter $y_1 = \cos x(\tan x - 1/\cos x)$ and $y_2 = \sin x - 1$. Note that we entered $\sec x$ as $1/\cos x$. For y_2 we will select the path graph style from the Style menu. To do this, highlight the expression for y_2 and then press $\boxed{\text{2nd}}$ $\boxed{\text{F6}}$ 6 or press $\boxed{\text{2nd}}$ $\boxed{\text{F6}}$ $\boxed{\triangledown}$ $\boxed{\triangledown}$ $\boxed{\triangledown}$ $\boxed{\triangledown}$ $\boxed{\triangledown}$ $\boxed{\text{ENTER}}$.

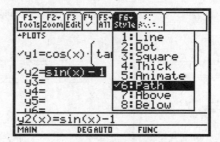

The calculator will graph y_1 first as a solid line. Then y_2 will be graphed as the circular cursor traces the leading edge of the graph, allowing us to determine visually whether the graphs coincide. In this case, the graphs appear to coincide, so the factorization is probably correct.

FINDING INVERSE FUNCTION VALUES

We can use a graphing calculator to find inverse function values in both radians and degrees.

Section 7.4, Example 2 (a), (e) Approximate $\cos^{-1}(-0.2689)$ and $\csc^{-1} 8.205$ in both radians and degrees.

To find inverse function values in radians, first select Radian for the Angle mode. Then, to approximate $\cos^{-1}(-0.2689)$ press $\boxed{\diamond}$ $\boxed{\text{COS}^{-1}}$ $\boxed{(-)}$ $\boxed{\cdot}$ 2 6 8 9 $\boxed{)}$ $\boxed{\text{ENTER}}$. (COS^{-1} is the green $\diamond$ operation associated with the $\boxed{\text{Z}}$ key.) The calculator returns 1.84304711, so $\cos^{-1}(-0.2689) \approx 1.8430$ radians. We also use reciprocal relationships to find function values for arcsecant and arccotangent.

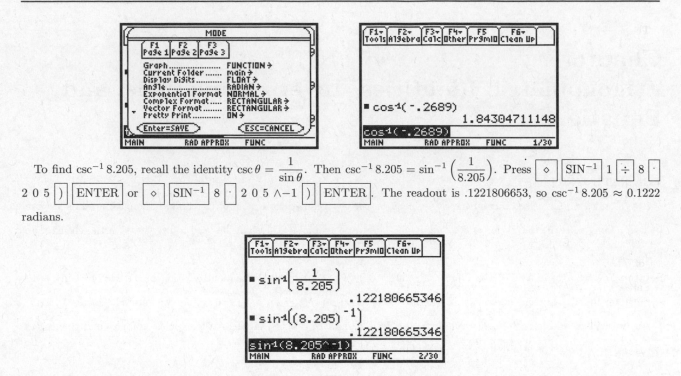

To find $\csc^{-1} 8.205$, recall the identity $\csc\theta = \dfrac{1}{\sin\theta}$. Then $\csc^{-1} 8.205 = \sin^{-1}\left(\dfrac{1}{8.205}\right)$. Press $\boxed{\diamond}$ $\boxed{\text{SIN}^{-1}}$ 1 $\boxed{\div}$ 8 $\cdot$ $2\ 0\ 5$ $\boxed{)}$ $\boxed{\text{ENTER}}$ or $\boxed{\diamond}$ $\boxed{\text{SIN}^{-1}}$ 8 $\cdot$ $2\ 0\ 5$ $\wedge{-}1$ $\boxed{)}$ $\boxed{\text{ENTER}}$. The readout is .1221806653, so $\csc^{-1} 8.205 \approx 0.1222$ radians.

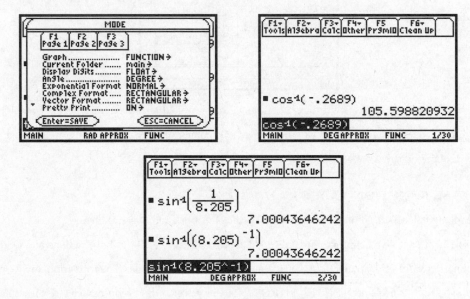

To find inverse function values in degrees, set the calculator in degree mode. Then use the keystrokes above to find that $\cos^{-1}(-0.2689) \approx 105.6°$ and $\csc^{-1} 8.205 \approx 7.0°$.

SINE REGRESSION

The SinReg operation can be used to fit a sine curve $y = a\sin(bx + c) + d$ to a set of data. At least four data points are required and there must be at least two data points per period. The output of SinReg is always in radians, regardless of the Radian/Degree mode setting. To see the graph, however, we must set the calculator in Radian mode.

The operations of entering data, making scatterplots, and graphing and evaluating the regression function are the same as for linear regression functions. Reread the material on pages 91 - 94 of this manual to review these procedures.

Section 7.5, Exercise 51 (a) Sales of certain products fluctuate in cycles. The data in the following table show the total sales of skis per month for a business in a northern climate.

Month, x	Total Sales, y (in thousands)
August, 8	$ 0
November, 11	7
February, 2	14
May, 5	7
August, 8	0

Using the sine regression feature on a graphing calculator, fit a sine function of the form $y = A\sin(Bx - C) + D$ to this set of data.

Enter the data in the Data/Matrix editor. Then press F5 to display the Calculate menu. Select SinReg by pressing ▷ alpha B . Then press ▽ alpha C 1 ▽ alpha C 2 to indicate that the data in c1 and c2 will be used for x and y, respectively. Press ▽ ▷ ▽ ENTER to indicate that the regression equation should be copied to the equation-editor screen as y_1. Finally press ENTER again to see the STAT VARS screen which displays the coefficients a, b, c, and d of the sine regression function $y = a\sin(bx + c) + d$

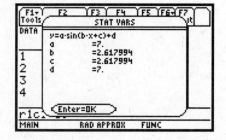

Chapter 8
Applications of Trigonometry

FINDING TRIGONOMETRIC NOTATION FOR COMPLEX NUMBERS

The TI-89 can be used to find trigonometric notation for a complex number.

Section 8.3, Example 3 (a) Find trigonometric notation for $1 + i$.

Trigonometric notation for a complex number has the form $r(\cos\theta + i\sin\theta)$. We can find r using the abs feature from the MATH Number menu. Press $\boxed{\text{2nd}}$ $\boxed{\text{MATH}}$ $\boxed{\triangleright}$ to display this menu. Then press 2 to copy "abs" to the home screen. (We could also use $\boxed{\triangledown}$ to highlight 2 and then press $\boxed{\text{ENTER}}$.) Then press 1 $\boxed{+}$ $\boxed{\text{2nd}}$ $\boxed{i}$ $\boxed{)}$ $\boxed{\text{ENTER}}$. The calculator returns $|1+i|$, the value of r. When Auto or Exact is selected for the Exact/Approx mode setting the calculator returns the exact value of r, $\sqrt{2}$. When Approximate is selected the calculator returns a decimal approximation for $\sqrt{2}$, 1.414213562.

Now use the MATH Complex menu again to find θ. We will do this first with Degree selected for the Angle mode. Press $\boxed{\text{2nd}}$ $\boxed{\text{MATH}}$ 5 to display the MATH Complex menu. Select item 4, "angle," by pressing 4 or by using $\boxed{\triangledown}$ to highlight 4 and then pressing $\boxed{\text{ENTER}}$. Then press 1 $\boxed{+}$ $\boxed{\text{2nd}}$ $\boxed{i}$ $\boxed{)}$ $\boxed{\text{ENTER}}$. The calculator returns 45, so the angle θ is 45°. When Radian and Auto or Exact modes are selected the calculator returns $\pi/4$.

CONVERTING FROM RECTANGULAR TO POLAR COORDINATES

A graphing calculator can be used to convert from rectangular to polar coordinates, expressing the result using either degrees or radians. The calculator will supply a positive value for r and an angle in the interval $(-180°, 180°]$, or $(-\pi, \pi]$.

Section 8.4, Example 2 (a) Convert (3,3) to polar coordinates.

To find r, regardless of the type of angle measure, press $\boxed{\text{2nd}}$ $\boxed{\text{MATH}}$ 2 5 3 $\boxed{,}$ 3 $\boxed{)}$ $\boxed{\text{ENTER}}$. When the calculator is set in Auto or Exact mode, it returns $3\sqrt{2}$. Now, to find θ in degrees, set the calculator in Degree mode and press $\boxed{\text{2nd}}$ $\boxed{\text{MATH}}$ 2 6 3 $\boxed{,}$ 3 $\boxed{)}$ $\boxed{\text{ENTER}}$. The readout is 45, so $\theta = 45°$. Thus polar notation for (3,3) is $(3\sqrt{2}, 45°)$.

Set the calculator in Radian mode to find θ in radians. Repeat the keystrokes for finding θ above to find that $\theta = \pi/4$. Thus polar notation for (3,3) is $(3\sqrt{2}, \pi/4)$.

```
┌─────────────────────────────────────┐
│ F1▾  F2▾  F3▾ F4▾  F5    F6▾         │
│Tools Algebra Calc Other Pr9miO Clean Up│
│                                     │
│                                     │
│ ■ R▸Pr(3, 3)                   3·√2 │
│ ■ R▸Pθ(3, 3)                     45 │
│                                   π │
│ ■ R▸Pθ(3, 3)                      ─ │
│                                   4 │
│ R▸Pθ(3,3)                           │
│ MAIN       RAD AUTO      FUNC   3/30 │
└─────────────────────────────────────┘
```

CONVERTING FROM POLAR TO RECTANGULAR COORDINATES

A graphing calculator can be used to convert from polar to rectangular coordinates.

Section 8.4, Example 3 Convert each of the following to rectangular coordinates.

(a) $(10, \pi/3)$ (b) $(-5, 135°)$

(a) Since the angle is given in radians, set the calculator in Radian mode. To find the x-coordinate of rectangular notation, press [2nd] [MATH] 2 3 1 0 [,] [2nd] [π] [÷] 3 [)] [ENTER]. The readout is 5, so $x = 5$. The y-coordinate is found by pressing [2nd] [MATH] 2 4 1 0 [,] [2nd] [π] [÷] 3 [)] [ENTER]. When the calculator is set in Auto or Exact mode, it returns $5\sqrt{3}$. Thus, rectangular notation for $(10, \pi/3)$ is $(5, 5\sqrt{3})$.

```
┌─────────────────────────────────────┐
│ F1▾  F2▾  F3▾ F4▾  F5    F6▾         │
│Tools Algebra Calc Other Pr9miO Clean Up│
│                                     │
│                                     │
│        ⎛    π⎞                      │
│ ■ P▸Rx⎜10, ─⎟                    5  │
│        ⎝    3⎠                      │
│        ⎛    π⎞                      │
│ ■ P▸Ry⎜10, ─⎟                 5·√3  │
│        ⎝    3⎠                      │
│ P▸Ry(10,π/3)                        │
│ MAIN       RAD AUTO      FUNC   2/30 │
└─────────────────────────────────────┘
```

(b) The angle is given in degrees, so we set the calculator in Degree mode. To find the x-coordinate of rectangular notation, press [2nd] [MATH] 2 3 [(−)] 5 [,] 1 3 5 [)] [ENTER]. When the calculator is set in Auto or Exact mode, it returns $\frac{5\sqrt{2}}{2}$. The y-coordinate is found by pressing [2nd] [MATH] 2 4 [(−)] 5 [,] 1 3 5 [)] [ENTER]. The readout is $-\frac{5\sqrt{2}}{2}$. Thus, rectangular notation for $(-5, 135°)$ is $(\frac{5\sqrt{2}}{2}, -\frac{5\sqrt{2}}{2})$.

```
┌─────────────────────────────────────┐
│ F1▾  F2▾  F3▾ F4▾  F5    F6▾         │
│Tools Algebra Calc Other Pr9miO Clean Up│
│                                     │
│                                     │
│                                5·√2 │
│ ■ P▸Rx(-5, 135)                ──── │
│                                  2  │
│                               -5·√2 │
│ ■ P▸Ry(-5, 135)               ──── │
│                                  2  │
│ P▸Ry(-5,135)                        │
│ MAIN       DEG AUTO      FUNC   2/30 │
└─────────────────────────────────────┘
```

GRAPHING POLAR EQUATIONS

Polar equations can be graphed in either Radian mode or Degree mode. The equation must be written in the form $r = f(\theta)$ and the calculator must be set in Polar mode. Typically we begin with a range of $[0, 2\pi]$ or $[0°, 360°]$, but it might be necessary to increase the range to ensure that sufficient points are plotted to display the entire graph.

Section 8.4, Example 6 Graph: $r = 1 - \sin\theta$.

First set the calculator in Polar mode by pressing $\boxed{\text{MODE}}$ $\boxed{\triangleright}$ 3 $\boxed{\text{ENTER}}$ or $\boxed{\text{MODE}}$ $\boxed{\triangleright}$ $\boxed{\triangledown}$ $\boxed{\triangledown}$ $\boxed{\text{ENTER}}$ $\boxed{\text{ENTER}}$. We will also select Radian mode.

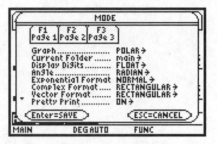

The equation is given in $r = f(\theta)$ form. Press $\boxed{\diamond}$ $\boxed{\text{Y} =}$ to go to the "Y =" screen. Clear any existing entries and, with the cursor beside "$r_1 =$," press 1 $\boxed{-}$ $\boxed{\text{2nd}}$ $\boxed{\text{SIN}}$ $\boxed{\diamond}$ $\boxed{\theta}$ $\boxed{)}$ $\boxed{\text{ENTER}}$. (θ is the green $\diamond$ operation associated with the $\boxed{\wedge}$ key.) Now press $\boxed{\diamond}$ $\boxed{\text{WINDOW}}$ and enter the following settings:

$\theta\text{min} = 0$	(Smallest value of θ to be evaluated)
$\theta\text{max} = 2\pi$	(Largest value of θ to be evaluated)
$\theta\text{step} = \pi/24$	(Increment in θ values)
$x\text{min} = -4$	
$x\text{max} = 4$	
$x\text{scl} = 1$	
$y\text{min} = -3$	
$y\text{max} = 1$	
$y\text{scl} = 1$	

With these settings the calculator evaluates the function from $\theta = 0$ to $\theta = 2\pi$ in increments of $\pi/24$ and displays the graph in the square window $[-4, 4, -3, 1]$. Values entered in terms of π appear on the window screen as decimal approximations. Press $\boxed{\diamond}$ $\boxed{\text{GRAPH}}$ to display the graph.

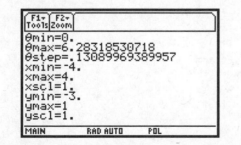 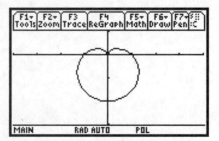

The curve can be traced with either rectangular or polar coordinates being displayed. The value of θ is also displayed when rectangular coordinates are selected. The choice of coordinates is made on the **GRAPH FORMATS** screen. While the graph is displayed, press $\boxed{\text{F1}}$ 9 to display this screen. Then press $\boxed{\triangleright}$ and select either RECT or POLAR.

Chapter 9
Systems of Equations and Matrices

QUADRATIC REGRESSION

Quadratic functions can be fit to data using the quadratic regression operation. The operations of entering data, making scatterplots, and graphing and evaluating quadratic regression functions are the same as for linear regression functions.

Section 9.2, Exercise 37 *Morning Newspapers* The number of morning newspapers in the United States in various years is shown in the following table.

Year	Number of Morning Newspapers
1920	437
1940	380
1960	312
1980	387
2000	766
2004	813

(a) Use a graphing calculator to fit a quadratic function to the data, where x is the number of years after 1920.

(b) Use the function found in part (a) to estimate the number of morning newspapers in 2008.

(a) Clear any existing entries on the equation-editor screen. Then enter the data in the Data/Matrix editor as described on page 92 of this manual. To fit a quadratic function to the data, press $\boxed{\text{APPS}}$ $\boxed{6}$ $\boxed{\text{ENTER}}$ $\boxed{\text{F5}}$ to view the Calculate menu. Then select QuadReg by pressing $\boxed{\triangleright}$ $\boxed{9}$. Then press $\boxed{\triangledown}$ $\boxed{\text{alpha}}$ $\boxed{\text{C}}$ $\boxed{1}$ $\boxed{\triangledown}$ $\boxed{\text{alpha}}$ $\boxed{\text{C}}$ $\boxed{2}$ to indicate that the data in c1 and c2 will be used for x and y, respectively. Press $\boxed{\triangledown}$ $\boxed{\triangleright}$ $\boxed{\triangledown}$ $\boxed{\text{ENTER}}$ to indicate that the regression equation should be copied to the equation-editor screen as y_1. Finally press $\boxed{\text{ENTER}}$ again to see the STAT VARS screen which displays the coefficients a, b, and c of the regression equation $y = ax^2 + bx + c$. Note that at least three data points are required for quadratic regression.

(b) To estimate the number of morning newspapers in 2008, we evaluate the regression function for 88. We can use any of the methods for evaluating a function found on pages 89 - 91 of this manual. Here we show a table set in Ask mode. We estimate that there will be about 892 morning newspapers in 2008.

MATRICES AND ROW-EQUIVALENT OPERATIONS

Matrices with up to 999 rows and 99 columns can be entered on a TI-89. Row-equivalent operations can be performed on matrices on the calculator.

Section 9.3, Example 1 Solve the following system:

$$
\begin{aligned}
2x - y + 4z &= -3, \\
x - 2y - 10z &= -6, \\
3x + 4z &= 7.
\end{aligned}
$$

First we will enter the augmented matrix

$$
\begin{bmatrix}
2 & -1 & 4 & -3 \\
1 & -2 & -10 & -6 \\
3 & 0 & 4 & 7
\end{bmatrix}
$$

in the Data/Matrix editor. We will call the Matrix A. Press $\boxed{\text{APPS}}$ 6 3 $\boxed{\triangleright}$ 2 $\boxed{\triangledown}$ $\boxed{\triangledown}$ $\boxed{\text{alpha}}$ $\boxed{\text{A}}$ $\boxed{\triangledown}$ 3 $\boxed{\triangledown}$ 4 $\boxed{\text{ENTER}}$ $\boxed{\text{ENTER}}$ to go to the Data/Matrix editor and set up a matrix named A with 3 rows and 4 columns. If a matrix named A has previously been saved in your calculator, an error message will be displayed. If this happens, you can select a different name for the matrix we are about to enter or you can delete the current matrix A and then enter the new matrix as A. To delete a matrix press $\boxed{\text{2nd}}$ $\boxed{\text{VAR-LINK}}$, use $\boxed{\triangledown}$ to highlight the name of the matrix being deleted, and then press $\boxed{\text{F1}}$ 1 $\boxed{\text{ENTER}}$ or $\boxed{\text{F1}}$ $\boxed{\text{ENTER}}$ $\boxed{\text{ENTER}}$. (VAR-LINK is the second operation associated with the $\boxed{-}$ key.)

Enter the elements of the first row of the matrix by pressing 2 $\boxed{\text{ENTER}}$ $\boxed{(-)}$ 1 $\boxed{\text{ENTER}}$ 4 $\boxed{\text{ENTER}}$ $\boxed{(-)}$ 3 $\boxed{\text{ENTER}}$. The cursor moves to the element in the second row and first column of the matrix. Enter the elements of the second and third rows of the augmented matrix by typing each in turn followed by $\boxed{\text{ENTER}}$ as above. Note that the screen displays only three columns of the matrix. The arrow keys can be used to move the cursor to any element at any time.

```
F1▼  F2        F3
Tools Plot Setup Cell        F6▼ F7
                            Util Stat
MAT
3x4   c1        c2        c3
1     2.        -1.       4.
2     1.        -2.       -10.
3     3.        0.        4.
4
r1c1=2.
MAIN        RAD APPROX    FUNC
```

Matrix operations are performed on the home screen and are found on the Math Matrix menu. Press $\boxed{\text{HOME}}$ or $\boxed{\text{2nd}}$ $\boxed{\text{QUIT}}$ to leave the matrix editor and go to this screen. Access the Math Matrix menu by pressing $\boxed{\text{2nd}}$ $\boxed{\text{MATH}}$ 4. (MATH is the second operation associated with the 5 numeric key.)

Press $\boxed{\text{alpha}}$ $\boxed{\text{J}}$ to see the four row-equivalent operations: rowSwap, rowAdd, nRow, and nRowAdd. These operations interchange two rows of a matrix, add two rows, multiply a row by a number, and multiply a row by a number and add it to a second row, respectively.

We will use the calculator to perform the row-equivalent operations that were done algebraically in the text. First, to interchange row 1 and row 2 of matrix **A**, with the MATH Matrix menu displayed, press 1 to select rowSwap. Then press $\boxed{\text{alpha}}$ $\boxed{\text{A}}$ to select **A**. Follow this with a comma and the rows to be interchanged, $\boxed{,}$ 1 $\boxed{,}$ 2 $\boxed{)}$ $\boxed{\text{ENTER}}$.

```
F1▼  F2▼   F3▼ F4▼  F5      F6▼
Tools Algebra Calc Other PrgmIO Clean Up

■ rowSwap(a,1,2)
        [1.  -2.  -10.  -6.]
        [2.  -1.   4.   -3.]
        [3.   0.   4.    7.]
rowSwap(a,1,2)
MAIN        RAD APPROX    FUNC
```

The calculator will not store the matrix produced using a row-equivalent operation, so when several operations are to be performed in succession it is helpful to store the result of each operation as it is produced. For example, to store the matrix resulting from interchanging the first and second rows of **A** as matrix **B** press $\boxed{\text{STO▷}}$ $\boxed{\text{alpha}}$ $\boxed{\text{B}}$ $\boxed{\text{ENTER}}$ immediately after interchanging the rows. This can also be done before $\boxed{\text{ENTER}}$ is pressed at the end of the rowSwap.

Next we multiply the first row of **B** by -2, add it to the second row and store the result as **B** again by pressing $\boxed{\text{2nd}}$ $\boxed{\text{MATH}}$ 4 $\boxed{\text{alpha}}$ $\boxed{\text{J}}$ 4 $\boxed{(-)}$ 2 $\boxed{,}$ $\boxed{\text{alpha}}$ $\boxed{\text{B}}$ $\boxed{,}$ 1 $\boxed{,}$ 2 $\boxed{)}$ $\boxed{\text{STO▷}}$ $\boxed{\text{alpha}}$ $\boxed{\text{B}}$ $\boxed{\text{ENTER}}$. These keystrokes select mRowAdd(from the MATH Matrix menu; then they specify that the value of the multiplier is -2, the matrix being operated on is **B**, and that a multiple of row 1 is being added to row 2; finally they store the result as **B**.

To multiply row 1 by -3, add it to row 3, and store the result as **B** press $\boxed{\text{2nd}}$ $\boxed{\text{MATH}}$ 4 $\boxed{\text{alpha}}$ $\boxed{\text{J}}$ 4 $\boxed{(-)}$ 3 $\boxed{,}$ $\boxed{\text{alpha}}$ $\boxed{\text{B}}$ $\boxed{,}$ 1 $\boxed{,}$ 3 $\boxed{)}$ $\boxed{\text{STO▷}}$ $\boxed{\text{alpha}}$ $\boxed{\text{B}}$ $\boxed{\text{ENTER}}$.

Now multiply the second row by 1/3 and store the result as **B** again. Press [2nd] [MATH] 4 [alpha] [J] 3 1 [(÷)] 3 , [alpha] [B] , 2) [STO▷] [alpha] [B] [ENTER]. These keystrokes select mRow(from the MATH Matrix menu; then they specify that the value of the multiplier is 1/3, the matrix being operated on is **B**, and row 2 is being multiplied; finally they store the result as **B**.

Multiply the second row by −6 and add it to the third row using mRowAdd(. Press [2nd] [MATH] 4 [ALPHA] [J] 4 [(−)] 6 , [alpha] [B] , 2 , 3) [STO▷] [alpha] [B] [ENTER]. The entry in the third row, second column is 1E−13. This is an approximation of 0 that occurs because of the manner in which the calculator performs calculations and should be treated as 0. In fact, it would be a good idea to return to the Data/Matrix editor at this point to replace this entry of **B** with 0. Press [APPS] 6 2 [▷] 2 [▽] [▽] [▷]. Then highlight **B** and press [ENTER] [ENTER]. Now highlight the entry in the third, row, second column and press 0 [ENTER]. Press [HOME] [alpha] [B] [ENTER] to see the edited version of **B**.

Finally, multiply the third row by −1/14 by pressing [2nd] [MATH] 4 [ALPHA] [J] 3 [(−)] 1 [(÷)] 1 4 , [alpha] [B] , 3) [ENTER].

```
/F1τ\/F2τ\/F3τ\/F4τ\/ F5 \/ F6τ  \
Tools│Algebra│Calc│Other│PrgmIO│Clean Up│
```

■ mRow(- 1/14, b, 3)

$$\begin{bmatrix} 1. & -2. & -10. & -6. \\ 0. & 1. & 8. & 3. \\ 0. & 0. & -1. & -.5 \end{bmatrix}$$

```
mRow( -1/14,b,3)
MAIN          RAD APPROX     FUNC      1/30
```

Write the system of equations that corresponds to the final matrix. Then use back-substitution to solve for x, y, and z as illustrated in the text.

Instead of stopping with row-echelon form as we did above, we can continue to apply row-equivalent operations until the matrix is in reduced row-echelon form as in **Example 3** in **Section 9.3** of the text. Reduced row-echelon form of a matrix can be found directly by using the rref(operation from the MATH Matrix menu. For example, to find reduced row-echelon form for matrix **A** in Example 1 above, after entering **A** and leaving the Data/Matrix screen $\boxed{\text{2nd}}$ $\boxed{\text{MATH}}$ 4 4 $\boxed{\text{alpha}}$ $\boxed{\text{A}}$ $\boxed{)}$ $\boxed{\text{ENTER}}$. We can read the solution of the system of equations, $(3, 7, -0.5)$ directly from the resulting matrix.

```
/F1τ\/F2τ\/F3τ\/F4τ\/ F5 \/ F6τ  \
Tools│Algebra│Calc│Other│PrgmIO│Clean Up│
```

■ rref(a)

$$\begin{bmatrix} 1. & 0. & 0. & 3. \\ 0. & 1. & 0. & 7. \\ 0. & 0. & 1. & -.5 \end{bmatrix}$$

```
rref(a)
MAIN          RAD APPROX     FUNC      1/30
```

MATRIX OPERATIONS

We can use a graphing calculator to add and subtract matrices, to multiply a matrix by a scalar, and to multiply matrices.

Section 9.4, Example 1 (a) Find **A** + **B** for

a) $\mathbf{A} = \begin{bmatrix} -5 & 0 \\ 4 & \frac{1}{2} \end{bmatrix}$, $\mathbf{B} = \begin{bmatrix} 6 & -3 \\ 2 & 3 \end{bmatrix}$.

Enter **A** and **B** in the Data/Matrix editor as described earlier in this chapter of the Graphing Calculator Manual. If you used the matrix names **A** and **B** above in Example 1 from Section 9.3, it will be necessary to delete those matrices before matrices **A** and **B** from this example can be entered. See page 126 of this manual for the procedure. Press $\boxed{\text{HOME}}$ or $\boxed{\text{2nd}}$ $\boxed{\text{QUIT}}$ to leave the Data/Matrix screen. Then press $\boxed{\text{alpha}}$ $\boxed{\text{A}}$ $\boxed{+}$ $\boxed{\text{alpha}}$ $\boxed{\text{B}}$ $\boxed{\text{ENTER}}$ to display the sum.

```
/F1τ\/F2τ\/F3τ\/F4τ\/ F5 \/ F6τ  \
Tools│Algebra│Calc│Other│PrgmIO│Clean Up│
```

■ a + b

$$\begin{bmatrix} 1. & -3. \\ 6. & 3.5 \end{bmatrix}$$

```
a+b
MAIN          RAD APPROX     FUNC      1/30
```

Section 9.4, Example 2 Find $\mathbf{C} - \mathbf{D}$ for each of the following.

a) $\mathbf{C} = \begin{bmatrix} 1 & 2 \\ -2 & 0 \\ -3 & -1 \end{bmatrix}$, $\mathbf{D} = \begin{bmatrix} 1 & -1 \\ 1 & 3 \\ 2 & 3 \end{bmatrix}$ b) $\mathbf{C} = \begin{bmatrix} 5 & -6 \\ -3 & 4 \end{bmatrix}$, $\mathbf{D} = \begin{bmatrix} -4 \\ 1 \end{bmatrix}$

a) Enter $\mathbf{C}$ and $\mathbf{D}$ in the Data/Matrix editor. Press $\boxed{\text{HOME}}$ or $\boxed{\text{2nd}}$ $\boxed{\text{QUIT}}$ to leave this screen. Then press $\boxed{\text{alpha}}$ $\boxed{\text{C}}$ $\boxed{-}$ $\boxed{\text{alpha}}$ $\boxed{\text{D}}$ $\boxed{\text{ENTER}}$ to display the difference.

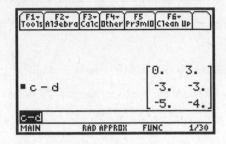

b) Enter $\mathbf{C}$ and $\mathbf{D}$ in the Data/Matrix editor after deleting $\mathbf{C}$ and $\mathbf{D}$ from part(a) above if necessary. (See page 126 of this manual for the procedure.) Press $\boxed{\text{HOME}}$ or $\boxed{\text{2nd}}$ $\boxed{\text{QUIT}}$ to leave the Data/Matrix screen. Then press $\boxed{\text{alpha}}$ $\boxed{\text{C}}$ $\boxed{-}$ $\boxed{\text{alpha}}$ $\boxed{\text{D}}$ $\boxed{\text{ENTER}}$. The calculator returns the error message "Dimension mismatch," indicating that this subtraction is not possible. This is the case because the matrices have different orders.

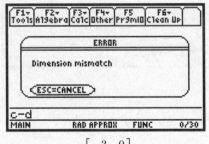

Section 9.4, Example 4 Find $3\mathbf{A}$ and $(-1)\mathbf{A}$, for $\mathbf{A} = \begin{bmatrix} -3 & 0 \\ 4 & 5 \end{bmatrix}$

Enter $\mathbf{A}$ in the Data/Matrix editor after deleting the matrix that was previously entered as $\mathbf{A}$ if necessary. (See page 126 of this manual for the procedure.) Press $\boxed{\text{HOME}}$ or $\boxed{\text{2nd}}$ $\boxed{\text{QUIT}}$ to leave the Data/Matrix screen. Then to find $3\mathbf{A}$ press 3 $\boxed{\text{alpha}}$ $\boxed{\text{A}}$ $\boxed{\text{ENTER}}$ and to find $(-1)\mathbf{A}$ press $\boxed{(-)}$ 1 $\boxed{\text{alpha}}$ $\boxed{\text{A}}$ $\boxed{\text{ENTER}}$. Note that $(-1)\mathbf{A}$ is the opposite, or additive inverse, of $\mathbf{A}$ and can also be found by pressing $\boxed{(-)}$ $\boxed{\text{alpha}}$ $\boxed{\text{A}}$ $\boxed{\text{ENTER}}$.

Section 9.4, Example 6 (a), (d) For

$$\mathbf{A} = \begin{bmatrix} 3 & 1 & -1 \\ 2 & 0 & 3 \end{bmatrix}, \mathbf{B} = \begin{bmatrix} 1 & 6 \\ 3 & -5 \\ -2 & 4 \end{bmatrix}, \text{ and } \mathbf{C} = \begin{bmatrix} 4 & -6 \\ 1 & 2 \end{bmatrix}$$

find each of the following.

a) **AB**   d) **AC**

First enter **A**, **B**, and **C** in the Data/Matrix editor after first deleting any matrices that were previously entered with those names. (See page 126 of this manual for the procedure.) Then press $\boxed{\text{HOME}}$ or $\boxed{\text{2nd}}$ $\boxed{\text{QUIT}}$ to leave this screen.

a) To find **AB** press $\boxed{\text{alpha}}$ $\boxed{\text{A}}$ $\boxed{\times}$ $\boxed{\text{alpha}}$ $\boxed{\text{B}}$ $\boxed{\text{ENTER}}$. Note that the multiplication symbol must be used so that the calculator can differentiate this multiplication from a variable named *ab*.

d) To find **AC** press $\boxed{\text{alpha}}$ $\boxed{\text{A}}$ $\boxed{\times}$ $\boxed{\text{alpha}}$ $\boxed{\text{C}}$ $\boxed{\text{ENTER}}$. The calculator returns the error message "Dimension," indicating that this multiplication is not possible. This is the case because the number of columns in **A** is not the same as the number of rows in **C**. Thus, the matrices cannot be multiplied in this order.

FINDING THE INVERSE OF A MATRIX

The inverse of a matrix can be found quickly on a graphing calculator.

Section 9.5, Example 3 Find $\mathbf{A}^{-1}$, where

$$\mathbf{A} = \begin{bmatrix} -2 & 3 \\ -3 & 4 \end{bmatrix}.$$

Enter **A** in the Data/Matrix editor after first deleting the matrix that was previously entered as **A** if necessary. (See page 126 of this manual for the procedure.) Then press $\boxed{\text{HOME}}$ or $\boxed{\text{2nd}}$ $\boxed{\text{QUIT}}$ to leave the Data Matrix screen. Now press $\boxed{\text{alpha}}$ $\boxed{\text{A}}$ $\boxed{\wedge}$ $\boxed{(-)}$ 1 $\boxed{\text{ENTER}}$ to display $\mathbf{A}^{-1}$.

Section 9.5, Exercise 7 Find $\mathbf{A}^{-1}$, where

$$\mathbf{A} = \begin{bmatrix} 6 & 9 \\ 4 & 6 \end{bmatrix}.$$

Enter $\mathbf{A}$ in the Data/Matrix editor after first deleting the matrix that was previously entered as $\mathbf{A}$ if necessary. (See page 126 of this manual for the procedure.) Then press $\boxed{\text{HOME}}$ or $\boxed{\text{2nd}}$ $\boxed{\text{QUIT}}$ to leave the Data Matrix screen. Now press $\boxed{\text{alpha}}$ $\boxed{\text{A}}$ $\boxed{\wedge}$ $\boxed{(-)}$ 1 $\boxed{\text{ENTER}}$. The calculator returns the message "Singular matrix," indicating that $\mathbf{A}^{-1}$ does not exist.

MATRIX SOLUTIONS OF SYSTEMS OF EQUATIONS

We can write a system of n linear equations in n variables as a matrix equation $\mathbf{AX} = \mathbf{B}$. If $\mathbf{A}$ has an inverse the solution of the system of equations is given by $\mathbf{X} = \mathbf{A}^{-1}\mathbf{B}$.

Section 9.5, Example 5 Use an inverse matrix to solve the following system of equations:

$$-2x + 3y = 4,$$
$$-3x + 4y = 5.$$

Enter $\mathbf{A} = \begin{bmatrix} -2 & 3 \\ -3 & 4 \end{bmatrix}$ and $\mathbf{B} = \begin{bmatrix} 4 \\ 5 \end{bmatrix}$ in the Data/Matrix editor after first deleting any matrices that were previously entered with those names. (See page 126 of this manual for the procedure.) Then press $\boxed{\text{HOME}}$ or $\boxed{\text{2nd}}$ $\boxed{\text{QUIT}}$ to leave the Data/Matrix screen. Press $\boxed{\text{alpha}}$ $\boxed{\text{A}}$ $\boxed{\wedge}$ $\boxed{(-)}$ 1 $\boxed{\times}$ $\boxed{\text{alpha}}$ $\boxed{\text{B}}$ $\boxed{\text{ENTER}}$. The result is the 2 x 1 matrix $\begin{bmatrix} 1 \\ 2 \end{bmatrix}$, so the solution is $(1, 2)$.

DETERMINANTS AND CRAMER'S RULE

We can evaluate determinants on a graphing calculator and use Cramer's rule to solve systems of equations.

Section 9.6, Example 5 Use a graphing calculator to evaluate $|\mathbf{A}|$.

$$\mathbf{A} = \begin{bmatrix} 1 & 6 & -1 \\ -3 & -5 & 3 \\ 0 & 4 & 2 \end{bmatrix}$$

Enter $\mathbf{A}$ in the Data/Matrix editor and then press $\boxed{\text{HOME}}$ or $\boxed{\text{2nd}}$ $\boxed{\text{QUIT}}$ to leave this screen. We will select the "det(" operation from the MATH Matrix menu. Press $\boxed{\text{2nd}}$ $\boxed{\text{MATH}}$ 4 2 $\boxed{\text{alpha}}$ $\boxed{\text{A}}$ $\boxed{)}$ $\boxed{\text{ENTER}}$. We see that $|\mathbf{A}| = 26$.

```
F1▾  F2▾  F3▾ F4▾   F5     F6▾
Tools Algebra Calc Other PrgmIO Clean Up




■ det(a)                              26
det(a)
MAIN       DEG AUTO      FUNC     1/30
```

Section 9.6, Example 6 Solve using Cramer's rule:

$$2x + 5y = 7,$$
$$5x - 2y = -3.$$

First we enter the matrices corresponding to D, D_x, and D_y as **A**, **B**, and **C**, respectively. We have

$$\mathbf{A} = \begin{bmatrix} 2 & 5 \\ 5 & -2 \end{bmatrix}, \mathbf{B} = \begin{bmatrix} 7 & 5 \\ -3 & -2 \end{bmatrix}, \text{ and } \mathbf{C} = \begin{bmatrix} 2 & 7 \\ 5 & -3 \end{bmatrix}.$$

Then we use the "det("operation from the MATRIX MATH menu. We have

$$x = \frac{\det(\mathbf{B})}{\det(\mathbf{A})} \text{ and } y = \frac{\det(\mathbf{C})}{\det(\mathbf{A})}.$$

To find x press $\boxed{\text{2nd}}$ $\boxed{\text{MATH}}$ $\boxed{4\ 2}$ $\boxed{\text{alpha}}$ $\boxed{\text{B}}$ $\boxed{)}$ $\boxed{\div}$ $\boxed{\text{2nd}}$ $\boxed{\text{MATH}}$ $\boxed{4\ 2}$ $\boxed{\text{alpha}}$ $\boxed{\text{A}}$ $\boxed{)}$ $\boxed{\text{ENTER}}$. The result is $-\dfrac{1}{29}$. To find y press $\boxed{\text{2nd}}$ $\boxed{\text{MATH}}$ $\boxed{4\ 2}$ $\boxed{\text{alpha}}$ $\boxed{\text{C}}$ $\boxed{)}$ $\boxed{\div}$ $\boxed{\text{2nd}}$ $\boxed{\text{MATH}}$ $\boxed{4\ 2}$ $\boxed{\text{alpha}}$ $\boxed{\text{A}}$ $\boxed{)}$ $\boxed{\text{ENTER}}$. We can also recall the entry for x and edit it, replacing matrix **B** with matrix **C**, to find y. In either case, the result is $\dfrac{41}{29}$. The solution of the system of equations is $\left(-\dfrac{1}{29}, \dfrac{41}{29}\right)$.

```
F1▾  F2▾  F3▾ F4▾   F5     F6▾
Tools Algebra Calc Other PrgmIO Clean Up


   det(b)
■ ───────                          - 1/29
   det(a)
   det(c)
■ ───────                           41/29
   det(a)
det(c)/det(a)
MAIN       DEG AUTO      FUNC     2/30
```

GRAPHS OF INEQUALITIES

We can graph linear inequalities on a graphing calculator, shading the region of the solution set. The calculator should be set in Func mode at this point.

Section 9.7, Example 1 Graph: $y < x + 3$.

First we graph the related equation $y = x + 3$. We use the standard window $[-10, 10, -10, 10]$. Since the inequality symbol is $<$ we know that the line $y = x + 3$ is not part of the solution set. In a hand-drawn graph we would use a dashed line to indicate this. After determining that the solution set of the inequality consists of all points below the line, we can use the "shade below" graph style to shade this region. After entering the related equation, highlight it and press $\boxed{\text{2nd}}$ $\boxed{\text{F6}}$ 8. Then press $\boxed{\diamond}$ $\boxed{\text{GRAPH}}$ to display the graph of the inequality. Keep in mind the fact that the line $y = x + 3$ is not included in the solution set.

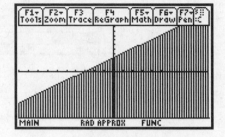

We can also use the Shade feature from the Catalog to graph this inequality. Copy Shade to the entry line of the home screen by pressing $\boxed{\text{HOME}}$ $\boxed{\text{CATALOG}}$ $\boxed{\text{S}}$, scrolling down to Shade, and pressing $\boxed{\text{ENTER}}$. We can also type Shade directly on the entry line by pressing $\boxed{\text{2nd}}$ $\boxed{\text{a-lock}}$ $\boxed{\text{S}}$ $\boxed{\text{H}}$ $\boxed{\text{A}}$ $\boxed{\text{D}}$ $\boxed{\text{E}}$ $\boxed{\text{alpha}}$.

Now enter a lower function and an upper function. The region between them will be shaded. We want to shade the area between the bottom of the window, $y = -10$, and the line $y = x + 3$ so we enter $\boxed{(-)}$ 1 0 $\boxed{\text{,}}$ $\boxed{\text{X}}$ $\boxed{\text{+}}$ 3 $\boxed{\text{ENTER}}$. We can also enter $x + 3$ as $y_1(x)$. The result is shown below. Keep in mind that the line $y = x + 3$ is not included in the solution set.

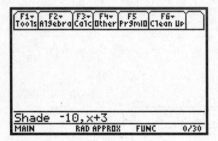 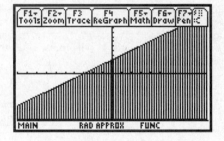

We can also graph a system of inequalities when the solution set lies between the graphs of two functions.

Section 9.7, Exercise 43 Graph:
$$y \leq x,$$
$$y \geq 3 - x.$$

First graph the related equations $y_1 = x$ and $y_2 = 3 - x$ and determine that the solution set consists of all the points on or below the graph of $y_1 = x$ and on or above the graph of $y_2 = 3 - x$. We can graph the system of inequalities by shading the solution set of each inequality in the system with a different pattern. When the "shade above" or "shade below" graph style options are selected the TI-89 rotates through four shading patterns. These patterns repeat if more than four functions are graphed. The region where the shaded areas overlap is the solution set of the inequality. Shade below $y_1 = x$ by highlighting the equation and then pressing $\boxed{\text{2nd}}$ $\boxed{\text{F6}}$ 8. Shade above $y_2 = 3 - x$ by highlighting the equation and pressing $\boxed{\text{2nd}}$ $\boxed{\text{F6}}$ 7. Then press $\boxed{\diamond}$ $\boxed{\text{GRAPH}}$.

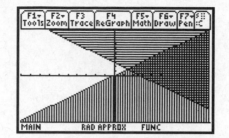

We can also use the Shade feature from the Catalog to graph this system of inequalities. Copy Shade to the entry line of the home screen or type it directly as described in Example 1 above. Then enter $3 - x$ as the lower function and x as the upper function.

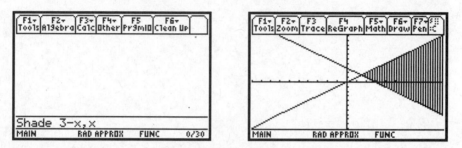

Chapter 10
Conic Sections

Many conic sections are represented by equations that are not functions. Consequently, these equations must be entered on the equation-editor screen of the TI-89 as two equations, each of which is a function.

GRAPHING PARABOLAS

To graph a parabola of the form $y^2 = 4px$ or $(y - k)^2 = 4p(x - h)$, we must first solve the equation for y.

Section 10.1, Example 4 Graph the parabola $y^2 - 2y - 8x - 31 = 0$.

In the text we used the quadratic formula to solve the equation for y:
$$y = \frac{2 \pm \sqrt{32x + 128}}{2}.$$

One way to produce the graph of the parabola is to enter $y_1 = \dfrac{2 + \sqrt{32x + 128}}{2}$ and $y_2 = \dfrac{2 - \sqrt{32x + 128}}{2}$, select a window, and press $\boxed{\diamond}$ $\boxed{\text{GRAPH}}$ to see the graph. Here we use $[-16, 16, -8, 8]$. The first equation produces the top half of the parabola and the second equation produces the lower half.

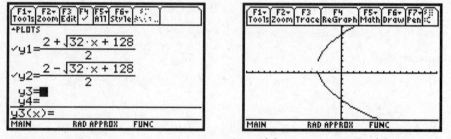

We can also enter $y_1 = \sqrt{32x + 128}$ and then enter $y_2 = \dfrac{2 + y_1(x)}{2}$ and $y_3 = \dfrac{2 - y_1(x)}{2}$. For example, to enter $y_2 = \dfrac{2 + y_1(x)}{2}$ position the cursor beside "$y_2 =$" and press $\boxed{(}$ $\boxed{2}$ $\boxed{+}$ $\boxed{Y}$ $\boxed{1}$ $\boxed{(}$ $\boxed{X}$ $\boxed{)}$ $\boxed{)}$ $\boxed{\div}$ 2. Enter $y_3 = \dfrac{2 - y_1}{2}$ in a similar manner. Finally, deselect y_1 by highlighting the expression for y_1 and pressing $\boxed{\text{F4}}$. Note that there is no longer a check mark beside y_1. This indicates that this equation has been deselected and thus its graph will not appear with the graph of the equations that remain selected. The top half of the graph is produced by y_2 and the lower half by y_3. The expression for y_1 was entered to avoid entering the square root more than once. By deselecting y_1 we prevent its graph from appearing on the screen with the graph of the parabola.

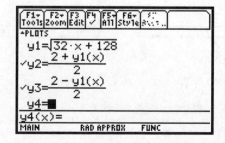

We could also use the standard equation of the parabola found in the text:

$$(y - 1)^2 = 8(x + 4).$$

Solve this equation for y.

$$y - 1 = \pm\sqrt{8(x+4)}$$
$$y = 1 \pm \sqrt{8(x+4)}$$

Then enter $y_1 = 1 + \sqrt{8(x+4)}$ and $y_2 = 1 - \sqrt{8(x+4)}$, or enter $y_1 = \sqrt{8(x+4)}$, $y_2 = 1 + y_1(x)$, and $y_3 = 1 - y_1(x)$, and deselect y_1 as described above.

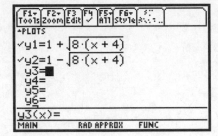

GRAPHING CIRCLES

The equation of an ellipse must be solved for y before it can be entered on the equation-editor screen of the TI-89. We can also use the Circle feature from the Catalog as described on page 88 of this manual.

Section 10.2, Example 1 Graph the circle $x^2 + y^2 - 16x + 14y + 32 = 0$.

In the text we found the standard form for the equation of the circle and then solved for y:

$$y = -7 \pm \sqrt{81 - (x-8)^2}.$$

We could also have solved the original equation using the quadratic formula.

One way to produce the graph is to enter $y_1 = -7 + \sqrt{81 - (x-8)^2}$ and $y_2 = -7 - \sqrt{81 - (x-8)^2}$, select a square window, and press $\boxed{\diamond}$ $\boxed{\text{GRAPH}}$. Here we use $[-16, 28, -18, 4]$. The first equation produces the top half of the circle and the second equation produces the lower half.

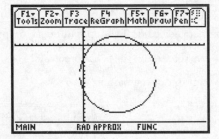

We can also enter $y_1 = \sqrt{81 - (x-8)^2}$ and then enter $y_2 = -7 + y_1(x)$ and $y_3 = -7 - y_1(x)$. Then deselect y_1, select a square window, and press $\boxed{\diamond}$ $\boxed{\text{GRAPH}}$. We use y_1 to eliminate the need to enter the square root more than once. Deselecting it prevents the graph of y_1 from appearing on the screen with the graph of the circle. The top half of the graph is produced by y_2 and the lower half by y_3.

GRAPHING ELLIPSES

The equation of an ellipse must be solved for y before it can be entered on the equation-editor screen of the TI-89. The procedure for graphing an ellipse of the form $\dfrac{x^2}{a^2} + \dfrac{y^2}{b^2} = 1$ or $\dfrac{x^2}{b^2} + \dfrac{y^2}{a^2} = 1$ is described on page 826 of the text. Here we consider ellipses of the form $\dfrac{(x-h)^2}{a^2} + \dfrac{(y-k)^2}{b^2} = 1$ or $\dfrac{(x-h)^2}{b^2} + \dfrac{(y-k)^2}{a^2} = 1$

Section 10.2, Example 4 Graph the ellipse $4x^2 + y^2 + 24x - 2y + 21 = 0$.

Completing the square in the text, we found that the equation can be written as

$$\frac{(x+3)^2}{4} + \frac{(y-1)^2}{16} = 1.$$

Solve this equation for y.

$$\frac{(x+3)^2}{4} + \frac{(y-1)^2}{16} = 1$$

$$\frac{(y-1)^2}{16} = 1 - \frac{(x+3)^2}{4}$$

$$(y-1)^2 = 16 - 4(x+3)^2 \qquad \text{Multiplying by 16}$$

$$y - 1 = \pm\sqrt{16 - 4(x+3)^2}$$

$$y = 1 \pm \sqrt{16 - 4(x+3)^2}$$

Now we can use this equation to produce the graph in either of two ways. One is to enter $y_1 = 1 + \sqrt{16 - 4(x+3)^2}$ and $y_2 = 1 - \sqrt{16 - 4(x+3)^2}$, select a square window, and press $\boxed{\diamond}$ $\boxed{\text{GRAPH}}$. Here we use $[-15, 9, -6, 6]$. The first equation produces the top half of the ellipse and the second equation produces the lower half.

We can also enter $y_1 = \sqrt{16 - 4(x+3)^2}$ and then enter $y_2 = 1 + y_1(x)$ and $y_3 = 1 - y_1(x)$. Deselect y_1, select a square window, and press $\boxed{\diamond}$ $\boxed{\text{GRAPH}}$. We use y_1 to eliminate the need to enter the square root more than once. Deselecting it prevents the graph of y_1 from appearing on the screen with the graph of the ellipse. The top half of the graph is produced by y_2 and the lower half by y_3.

```
┌─────────────────────────────────┐
│ F1▾  F2▾ F3 F4 F5▾ F6▾  ⋯        │
│ Tools Zoom Edit ✓ All Style ⋯    │
├─────────────────────────────────┤
│ ▲PLOTS                           │
│        ┌──────────────┐²         │
│  y1=√ 16 − 4·(x + 3)             │
│ ✓y2=1 + y1(x)                    │
│ ✓y3=1 − y1(x)                    │
│  y4=■                            │
│  y5=                             │
│  y6=                             │
├─────────────────────────────────┤
│ y4(x)=                           │
│ MAIN         RAD APPROX   FUNC   │
└─────────────────────────────────┘
```

We could also begin by using the quadratic formula to solve the original equation for y.

$$4x^2 + y^2 + 24x - 2y + 21 = 0$$
$$y^2 - 2y + (4x^2 + 24x + 21) = 0$$
$$y = \frac{-(-2) \pm \sqrt{(-2)^2 - 4 \cdot 1 \cdot (4x^2 + 24x + 21)}}{2 \cdot 1}$$
$$y = \frac{2 \pm \sqrt{4 - 16x^2 - 96x - 84}}{2}$$
$$y = \frac{2 \pm \sqrt{-16x^2 - 96x - 80}}{2}$$

Then enter $y_1 = \dfrac{2 + \sqrt{-16x^2 - 96x - 80}}{2}$ and $y_2 = \dfrac{2 - \sqrt{-16x^2 - 96x - 80}}{2}$, or enter $y_1 = \sqrt{-16x^2 - 96x - 80}$, $y_2 = \dfrac{2 + y_1(x)}{2}$, and $y_3 = \dfrac{2 - y_1(x)}{2}$, and deselect y_1.

```
┌─────────────────────────────────┐  ┌─────────────────────────────────┐
│ F1▾  F2▾ F3 F4 F5▾ F6▾  ⋯        │  │ F1▾  F2▾ F3 F4 F5▾ ⋯ ⋯           │
│ Tools Zoom Edit ✓ All Style ⋯    │  │ Tools Zoom Edit ✓ All ⋯ ⋯        │
├─────────────────────────────────┤  ├─────────────────────────────────┤
│ ▲PLOTS                           │  │ ▲DATA:main\ads                   │
│      2+√ −16·x²− 96·x − 80       │  │ ▮▮▮▮  ⊡ x:c1 y:c2                │
│ ✓y1=─────────────────────        │  │        ┌──────────────┐          │
│            2                     │  │  y1=√ −16·x²− 96·x − 80          │
│      2−√ −16·x²− 96·x − 80       │  │      2 + y1(x)                   │
│ ✓y2=─────────────────────        │  │ ✓y2=──────────                   │
│            2                     │  │          2                       │
│  y3=■                            │  │      2 − y1(x)                   │
├─────────────────────────────────┤  │ ✓y3=──────────                   │
│ y3(x)=                           │  │          2                       │
│ MAIN         RAD APPROX   FUNC   │  ├─────────────────────────────────┤
└─────────────────────────────────┘  │ MAIN         RAD APPROX   FUNC   │
                                      └─────────────────────────────────┘
```

Select a square window and press $\boxed{\diamond}$ $\boxed{\text{GRAPH}}$ to display the graph shown on the previous page.

GRAPHING HYPERBOLAS

As with equations of circles, parabolas, and ellipses, equations of hyperbolas must be solved for y before they can be entered on the equation-editor screen of the TI-89.

Section 10.3, Example 2 Graph the hyperbola $9x^2 - 16y^2 = 144$.

First solve the equation for y.
$$9x^2 - 16y^2 = 144$$
$$-16y^2 = -9x^2 + 144$$
$$y^2 = \frac{-9x^2 + 144}{-16}$$
$$y = \pm\sqrt{\frac{-9x^2 + 144}{-16}}, \text{ or } \pm\sqrt{\frac{9x^2 - 144}{16}}$$

It is not necessary to simplify further.

Now enter $y_1 = \sqrt{\dfrac{9x^2 - 144}{16}}$ and either $y_2 = -\sqrt{\dfrac{9x^2 - 144}{16}}$ or $y_2 = -y_1(x)$, select a square window, and press $\boxed{\diamond}$

GRAPH . Here we use the window $[-12, 12, -6, 6]$. The top half of the graph is produced by y_1 and the lower half by y_2.

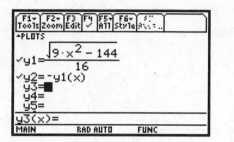 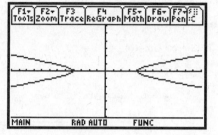

Section 10.3, Example 3 Graph the hyperbola $4y^2 - x^2 + 24y + 4x + 28 = 0$.

In the text we completed the square to get the standard form of the equation. Now solve the equation for y.

$$\frac{(y+3)^2}{1} - \frac{(x-2)^2}{4} = 1$$

$$(y+3)^2 = \frac{(x-2)^2}{4} + 1$$

$$y + 3 = \pm\sqrt{\frac{(x-2)^2}{4} + 1}$$

$$y = -3 \pm \sqrt{\frac{(x-2)^2}{4} + 1}$$

This equation can be used to produce the graph in either of two ways. One is to enter $y_1 = -3 + \sqrt{\frac{(x-2)^2}{4} + 1}$ and $y_2 = -3 - \sqrt{\frac{(x-2)^2}{4} + 1}$, select a square window, and press $\diamond$ GRAPH . Here we use $[-15, 15, -9, 6]$. The first equation produces the top half of the hyperbola and the second the lower half.

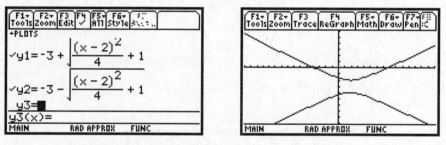

We can also enter $y_1 = \sqrt{\frac{(x-2)^2}{4} + 1}$, $y_2 = -3 + y_1(x)$, and $y_3 = -3 - y_1(x)$. Then deselect y_1, select a square window, and press $\diamond$ GRAPH . Again, y_1 is used to eliminate the need to enter the square root more than once. Deselecting it prevents the graph of y_1 from appearing on the screen with the graph of the hyperbola. The top half of the graph is produced by y_2 and the lower half by y_3.

GRAPHING PARAMETRIC EQUATIONS

Plane curves described with parametric equations can be graphed on a graphing calculator.

Section 10.7, Example 2(a) Using a graphing calculator, graph the plane curve given by the set of parametric equations and the restriction on the parameter.

$$x = t^2, \; y = t - 1,; \; -1 \le t \le 4$$

First press $\boxed{\text{MODE}}$ and select Parametric mode.

Then press $\boxed{\diamond}$ $\boxed{\text{Y} =}$ to display the equation-editor screen. Enter $xt1 = t^2$ and $yt1 = t - 1$. Now press $\boxed{\diamond}$ $\boxed{\text{WINDOW}}$ and enter the following settings:

tmin $= -1$	(Smallest value of t to be evaluated)
tmax $= 4$	(Largest value of t to be evaluated)
tstep $= .1$	(Increment in t values)
xmin $= -2$	
xmax $= 18$	
xscl $= 1$	
ymin $= -4$	
ymax $= 4$	
yscl $= 1$	

Since $x = t^2$ and $-1 \le t \le 4$, we have $0 \le x \le 16$. Thus, we choose xmin and xmax to display this interval. Similarly, since $y = t - 1$, we have $-2 \le y \le 3$ and we choose ymin and ymax to show this interval. Press $\boxed{\diamond}$ $\boxed{\text{GRAPH}}$ to display the graph.

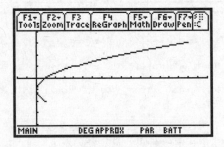

Chapter 11
Sequences, Series, and Combinatorics

The computational capabilities of the TI-89 can be used when working with sequences, series, and combinatorics.

FINDING THE TERMS OF A SEQUENCE

Section 11.1, Example 2 Use a graphing calculator to find the first 5 terms of the sequence whose general term is given by $a_n = n/(n+1)$.

Although we could use a table, we will use the Seq feature from the MATH List menu. Select Auto for the Exact/Approx mode setting so that the terms will be expressed in fractional form. Press $\boxed{\text{2nd}}$ $\boxed{\text{MATH}}$ 3 to display the MATH List menu. Then press 1 or $\boxed{\text{ENTER}}$ to paste "seq(" to the entry line of the home screen. Now enter the general term of the sequence. Follow this with the variable and the numbers of the first and last terms desired. Press $\boxed{\text{X}}$ $\boxed{\div}$ $\boxed{(}$ $\boxed{\text{X}}$ $\boxed{+}$ 1 $\boxed{)}$ $\boxed{,}$ $\boxed{\text{X}}$ $\boxed{,}$ 1 $\boxed{,}$ 5 $\boxed{)}$ $\boxed{\text{ENTER}}$.

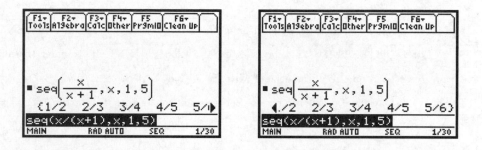

FINDING PARTIAL SUMS

We can use a TI-89 to find partial sums of a sequence when a formula for the general term is known.

Section 11.1, Example 6 Use a graphing calculator to find S_1, S_2, S_3, and S_4 for the sequence whose general term is given by $a_n = n^2 - 3$.

We will use the cumSum feature from the MATH List menu. The calculator will write the partial sums as a list. First press $\boxed{\text{2nd}}$ $\boxed{\text{MATH}}$ 3 7 to paste "cumSum(" to the home screen. Then press $\boxed{\text{2nd}}$ $\boxed{\text{MATH}}$ 3 $\boxed{\text{ENTER}}$ or $\boxed{\text{2nd}}$ $\boxed{\text{MATH}}$ 3 1 to paste "seq(" into the cumSum expression. Finally press $\boxed{\text{X}}$ $\boxed{\wedge}$ 2 $\boxed{-}$ 3 $\boxed{,}$ $\boxed{\text{X}}$ $\boxed{,}$ 1 $\boxed{,}$ 4 $\boxed{)}$ $\boxed{)}$ $\boxed{\text{ENTER}}$. We see that $S_1 = -2, S_3 = -1, S_3 = 5$, and $S_4 = 18$.

Section 11.1, Example 7 (a) Evaluate $\displaystyle\sum_{k=1}^{5} k^3$.

We will use the sum feature from the MATH List menu along with the seq feature. Press $\boxed{\text{2nd}}$ $\boxed{\text{MATH}}$ 3 6 $\boxed{\text{2nd}}$ $\boxed{\text{MATH}}$ 3 $\boxed{\text{ENTER}}$ $\boxed{\text{X}}$ $\boxed{\wedge}$ 3 $\boxed{,}$ $\boxed{\text{X}}$ $\boxed{,}$ 1 $\boxed{,}$ 5 $\boxed{)}$ $\boxed{)}$ $\boxed{\text{ENTER}}$.

RECURSIVELY DEFINED SEQUENCES

Recursively defined sequences can also be entered on a TI-89 set in Seq mode.

Section 11.1, Example 9 Find the first 5 terms of the sequence defined by

$$a_1 = 5, \ a_{k+1} = 2a_k - 3, \text{ for } k \ge 1.$$

Press $\boxed{\diamond}$ $\boxed{\text{Y}=}$ and enter the recursive function by positioning the cursor beside $u_1 =$ and pressing $\boxed{\text{ENTER}}$ 2 $\boxed{\text{alpha}}$ $\boxed{\text{U}}$ 1 $\boxed{(}$ $\boxed{\text{alpha}}$ $\boxed{\text{N}}$ $\boxed{-}$ 1 $\boxed{)}$ $\boxed{-}$ 3 $\boxed{\text{ENTER}}$. Then press $\boxed{\text{2nd}}$ $\boxed{\{}$ 5 $\boxed{\text{2nd}}$ $\boxed{\}}$ $\boxed{\text{ENTER}}$ to enter the initial value of 5 beside $ui_1 =$. ({ and } are the second operations associated with the $\boxed{(}$ and $\boxed{)}$ keys, respectively.)

Next press $\boxed{\diamond}$ $\boxed{\text{TblSet}}$ to display the Table Setup screen. Set Independent to Auto, tblStart = 1, Δtbl = 1. Also press $\boxed{\diamond}$ $\boxed{\text{WINDOW}}$ and check to be sure that nmin has the same value as tblStart. Then press $\boxed{\diamond}$ $\boxed{\text{TABLE}}$ to display the table of values for the recursive function. We see that $a_1 = 5$, $a_2 = 7$, $a_3 = 11$, $a_4 = 19$, and $a_5 = 35$.

EVALUATING FACTORIALS, PERMUTATIONS, AND COMBINATIONS

Operations from the MATH Probability menu can be used to evaluate factorials, permutations, and combinations.

Section 11.5, Exercise 6 Evaluate 7!.

On the home screen press 7 $\boxed{\text{2nd}}$ $\boxed{\text{MATH}}$ 7 1 $\boxed{\text{ENTER}}$ or 7 $\boxed{\text{2nd}}$ $\boxed{\text{MATH}}$ 7 $\boxed{\text{ENTER}}$ $\boxed{\text{ENTER}}$. These keystrokes enter 7, display the MATH Probability menu, select item 1, !, from that menu, and then cause 7! to be evaluated. The result is 5040.

Section 11.5, Exercise 9 Evaluate $\frac{9!}{5!}$.

Press 9 [2nd] [MATH] 7 1 [÷] 5 [2nd] [MATH] 7 1 [ENTER]. Both 1's can be replaced by [ENTER] if desired. The result is 3024.

```
┌─────────────────────────────────────┐
│ F1▼  F2▼   F3▼ F4▼  F5    F6▼        │
│Tools│Algebra│Calc│Other│PrgmIO│Clean Up│
├─────────────────────────────────────┤
│                                     │
│                                     │
│ ■ 7!                         5040   │
│   9!                                │
│ ■ ──                         3024   │
│   5!                                │
│ 9!/5!                               │
│MAIN      RAD AUTO     SEQ    2/30   │
└─────────────────────────────────────┘
```

Section 11.5, Example 3 (a) Compute $_4P_4$.

Press [2nd] [MATH] 7 2 4 [,] 4 [)] [ENTER]. These keystrokes display the MATH Probability menu, select item 2, $_nP_r$, from that menu, enter 4 for 4 objects and 4 for 4 objects taken at a time, and then cause the calculation to be performed. The result is 24.

Section 11.5, Example 6 Compute $_8P_4$.

Press [2nd] [MATH] 7 2 8 [,] 4 [)] [ENTER]. These keystrokes display the MATH Probability menu, select item 2, $_nP_r$, from that menu, enter 8 for 8 objects and 4 for 4 objects taken at a time, and then cause the calculation to be performed. The result is 1680. The previous entry could also be edited to obtain this result.

```
┌─────────────────────────────────────┐
│ F1▼  F2▼   F3▼ F4▼  F5    F6▼        │
│Tools│Algebra│Calc│Other│PrgmIO│Clean Up│
├─────────────────────────────────────┤
│                                     │
│                                     │
│ ■ nPr(4,4)                     24   │
│ ■ nPr(8,4)                   1680   │
│ nPr(8,4)                            │
│MAIN      RAD AUTO     SEQ    2/30   │
└─────────────────────────────────────┘
```

Section 11.6, Example 2 Evaluate $\binom{7}{5}$.

Press [2nd] [MATH] 7 3 7 [,] 5 [)] [ENTER]. These keystrokes display the MATH Probability menu, select item 3, $_nC_r$, from that menu, enter 7 for 7 objects and 5 for 5 objects taken at a time, and then cause the calculation to be performed. The result is 21.

```
┌─────────────────────────────────────┐
│ F1▼  F2▼   F3▼ F4▼  F5    F6▼        │
│Tools│Algebra│Calc│Other│PrgmIO│Clean Up│
├─────────────────────────────────────┤
│                                     │
│                                     │
│                                     │
│ ■ nCr(7,5)                     21   │
│ nCr(7,5)                            │
│MAIN      RAD AUTO     SEQ    1/30   │
└─────────────────────────────────────┘
```

Index
TI-83 Plus and TI-84 Plus Graphing Calculators

TI-83 Plus, TI-84 Plus Index

TI-83 Plus, TI-84 Plus Index

Index
TI-89 Graphing Calculator

TI-89 Graphing Calculator Index

TI-89 Graphing Calculator Index